M000232626

Anthropology and Development

Anthropology, Culture and Society

Series Editors:
Professor Vered Amit, Concordia University
and
Professor Christina Garsten, Stockholm University

Recent titles:

Anthropology and Development

Challenges for the Twenty-First Century

Katy Gardner and David Lewis

PlutoPress
www.plutobooks.com

First published 2015 by Pluto Press
345 Archway Road, London N6 5AA

www.plutobooks.com

British Library Cataloguing in Publication Data
A catalogue record for this book is available from the British Library

ISBN   978 0 7453 3365 6   Hardback
ISBN   978 0 7453 3364 9   Paperback
ISBN   978 1 7837 1275 5   PDF eBook
ISBN   978 1 7837 1277 9   Kindle eBook
ISBN   978 1 7837 1276 2   EPUB eBook

Library of Congress Cataloging in Publication Data applied for

10   9   8   7   6   5   4   3   2   1

Typeset by Stanford DTP Services, Northampton, England
Text design by Melanie Patrick
Simultaneously printed digitally by CPI Antony Rowe, Chippenham, UK
and Edwards Bros in the United States of America

Contents

Series Preface

Anthropology is a discipline based upon in-depth ethnographic works that deal with wider theoretical issues in the context of particular, local conditions – to paraphrase an important volume from the series: *large issues* explored in *small places*. This series has a particular mission: to publish work that moves away from an old-style descriptive ethnography that is strongly area-studies oriented and offer genuine theoretical arguments that are of interest to a much wider readership, but which are nevertheless located and grounded in solid ethnographic research. If anthropology is to argue itself a place in the contemporary intellectual world, then it must surely be through such research.

We start from the question: 'What can this ethnographic material tell us about the bigger theoretical issues that concern the social sciences?' rather than 'What can these theoretical ideas tell us about the ethnographic context?' Put this way round, such work becomes *about* large issues, *set in* a (relatively) small place, rather than detailed description of a small place for its own sake. As Clifford Geertz once said, 'Anthropologists don't study villages; they study *in* villages.'

By place, we mean not only geographical locale, but also other types of 'place' – within political, economic, religious or other social systems. We therefore publish work based on ethnography within political and religious movements, occupational or class groups, among youth, development agencies and nationalist movements; but also work that is more thematically based – on kinship, landscape, the state, violence, corruption, the self. The series publishes four kinds of volume: ethnographic monographs; comparative texts; edited collections; and shorter, polemical essays.

We publish work from all traditions of anthropology, and all parts of the world, which combines theoretical debate with empirical evidence to demonstrate anthropology's unique position in contemporary scholarship and the contemporary world.

Professor Vered Amit
Professor Christina Garsten

Preface

We originally chose to write this book for two main reasons. The first was that during the early 1990s we could find no single text that engaged with the various histories, opinions and debates that had emerged during the relationships between development people and anthropologists. While Lucy Mair's book *Anthropology and Development* (1984) was an invaluable text, it had been written well before both subjects embarked on periods of intensive reflection informed by the rise of postmodernist theory. We therefore hoped to move debates forward in a manner that could engage researchers, students and practitioners. Reviewers were on the whole kind about the book, and this made us feel that our work had at least gone some way towards achieving our intention.

The second reason was more personal. By the middle of the 1990s both of us were interested in trying to make sense of our disparate experiences to date working as anthropologists, researchers and development practitioners – in the field, at universities and research institutes, behind desks in development agencies and within interdisciplinary consultancy teams. Katy Gardner and David Lewis had both studied social anthropology as a first degree in the early 1980s. Katy Gardner's PhD research involved fieldwork in a Bangladeshi migrant village. After completing her dissertation, she spent a year working for the British Overseas Development Administration (ODA) as an assistant social adviser. During this period she was involved in short visits to various projects in South Asia as well as administrative work in London. Since leaving the ODA Katy has worked as a full-time lecturer in anthropology at the Universities of Kent and Sussex (where she worked from 1993 to 2013, becoming Professor of Anthropology in 2009). Earlier in her career she was also involved in a range of consultancy work for both private and governmental agencies. She is the author of *Songs at the River's Edge: Stories from a Bangladeshi Village* (Virago, 1991); *Global Migrants, Local Lives: Travel and Transformation in Rural Bangladesh* (Oxford University Press, 1995) and *Age, Narrative and Migration: Life History and the Life Course amongst Bengali Elders in London* (Berg, 2002). In 2013 she became professor of anthropology in the Department of Anthropology at the London School of Economics and Political Science. Katy Gardner's most recent book is

Discordant Development: Global Capitalism and the Struggle for Connection in Bangladesh (Pluto, 2012). Katy has also written several novels, including *Losing Gemma* (Penguin, 2001).

David Lewis moved from anthropology into a more interdisciplinary study of development. After a postgraduate course in development studies, he completed a PhD in rural sociology, in which he studied the effects of rural technological change in a Bangladeshi village. A five-year period of freelance research and consultancy work followed, including working as a Research Associate at the Overseas Development Institute (ODI) in London. He undertook research and consultancy work for a number of government and non-governmental agencies in Bangladesh, India, Nepal, Sri Lanka and Albania before becoming a full-time academic in the Department of Social Policy at the London School of Economics and Political Science. David Lewis's books include *Technologies and Transactions: A Study of the Interaction between Agrarian Structure and New Technology in Bangladesh* (Centre for Social Studies, University of Dhaka, 1991); *Trading the Silver Seed: Local Knowledge and Market Moralities in Aquacultural Development* (Intermediate Technology Publications, 1996); *The Management of Non-Governmental Development Organizations* (Routledge, 2001, 2007 and 2014); and *Development Brokers and Translators: The Ethnography of Aid and Agencies* (Kumarian, 2006, co-edited with David Mosse). His most recent books include *Bangladesh: Politics, Economy and Civil Society* (Cambridge University Press, 2011) and *Popular Representations of Development: Insights from Novels, Films, Television and Social Media* (co-edited with D. Rodgers and M. Woolcock, Routledge, 2014).

We ended our original preface with a few words about our overall intentions. We felt that many prevailing assumptions about and approaches to development were flawed or basically wrong-headed, but we did not see much value in simply being critical without trying to offer any creative alternatives. Instead, we believed in options that were rooted in reality rather than simply in rhetoric, in breaking down the barriers which existed between 'developers' and the 'developed', and in the need for a fuller critical discussion about what 'development' meant, one which reflected a true multiplicity of voices. We believed that there was a pressing moral and political responsibility to work towards improving the quality of life for the bulk of the world's population, and that in general a poor job was being made of this task. We did not wish to suggest that anthropology could somehow 'save' the development industry, nor that it could necessarily make the process of change more benign. However, we believed that anthropologists and development practitioners might have something to learn from each other, in order that better futures may be imagined and, perhaps, brought into being. We have not changed our views on these issues.

In returning to this book almost two decades after we originally started work on the first version, we were faced with some difficult choices. Should we simply update each chapter to take account of the new anthropological work that has emerged since 1996 or, given the myriad ways in which both ourselves and the world have changed, completely rewrite the book? In the end we decided to substantially rewrite around three-quarters of the text. Some sections from the earlier book are retained but lightly edited, but most of the book is new and the title has been changed. We leave readers to decide whether this is a second edition or a new book.

Katy Gardner and David Lewis
September 2014

Note: In writing about some of these experiences as ethnography we have, for obvious reasons, disguised some particulars in our accounts in terms of places and organisations, in keeping with the anthropological tradition of preserving the anonymity of informants.

Acknowledgements

First time around we thanked Eric Worby, Dina Siddiqi, Ben Crow, Sushila Zeitlyn, B.K. Jahangir, S.M. Nurul Alam, Sue Phillips and Emma Crewe for their stimulating discussions about many of these issues and for their encouragement during the long period of writing. We were very grateful to Richard Wilson for commissioning the book and for providing useful editorial comments and support, to James Fairhead for reading the original manuscript and providing valuable insights, and to Hamish Arnott for help with proofreading. We dedicated the book to the memory of Jonathan Zeitlyn, whose open mind, personal warmth and commitment to working towards a fairer world continues to inspire both of us.

Second time around we are once again incredibly grateful to James Fairhead for reading and providing very useful comments on our draft. We'd also like to thank David Castle at Pluto for his faith in the book and for persuading us to prepare a new edition.

Katy Gardner and David Lewis

Glossary

Development jargon

accountability – making development interventions more responsive to the people they seek to assist; also used by donors to mean making sure that money is used for the purpose for which it was intended

applied anthropology – the application of anthropological research to solving practical problems in development, public health, administration, industry, etc.

appropriate technology – the idea of viewing technology in the context of people's needs, drawn originally from the work of E.F. Schumacher in the 1970s, in reaction to 'hi-tech' solutions to problems of poverty

basic needs – a development strategy devised in the 1970s by governments and UN agencies in reaction to disillusionment with 'trickle down'

beneficiaries – those people whom a development project is intended to assist

bottom-up – interventions which come from the grassroots as opposed to government planners or development agencies

civil society – a political science concept with a long and complex history, essentially referring to the idea of independent organised citizen action as complementing, opposing, or forming a counterweight to the power of the state and the market

community development – the attempt to strengthen the institutions of local communities in order that they will sustain the gains brought about by a development project

conditionality – the imposition of terms by an aid giver upon a government or an organisation receiving the assistance (e.g. a bilateral donor gives a loan to an NGO provided it is used to support particular activities)

donor – usually refers to government agencies such as the UK Overseas Development Administration (ODA) or United States Agency for International Development (USAID), or to multilateral agencies such as the World Bank, but also includes NGOs such as Oxfam who fund partner organisations in the countries where they work

empowerment – the transformative potential of people to achieve positive changes in their lives by asserting their rights as women, citizens, etc., usually by group action, and thereby gaining greater power to solve problems

evaluation – the task of assessing whether or not a development intervention has been successful in meeting its objectives

green revolution – the introduction of new agricultural technologies including high yielding hybrid seeds, mechanised irrigation equipment and chemical fertiliser that began in the 1960s

non-governmental organisation (NGO) – there are many types: international, national and local; large and small; specialised (e.g. health, agriculture) or general (combining many sectors of activity); membership or non-membership. NGOs are not-for-profit development organisations, many of which depend on donations from members, the public or development agencies. Many are contractors with governments. In the US, NGOs are often known as private voluntary organisations (PVOs)

the North – along with 'the South', the term originated recently as less pejorative alternatives to 'First World' and 'Third World'. But both terms continue to cause problems by insisting that poverty can be geographically specified

participation – used to describe greater involvement by 'beneficiaries' in deciding the type of development projects they need, and how they are run. The degree of this involvement can, however, vary greatly

project – an intervention aimed at promoting social change usually by, or with the support of, an outside agency for a finite period (anything from a few years to several decades)

social development – a new term used in the UK to describe the 'softer' elements of the development process as distinct from economic and technical issues – education, health-care, human rights, etc.

social movements – loosely organised groups around the world taking issue-based action in a variety of areas (human rights, environment, access to land, gender rights, peace, etc.) usually local, without outside assistance at least in the first instance

the South – see entry for 'the North'

structural adjustment – policies which became common during the 1980s, introduced by the World Bank, as conditionality on loans, aimed at improving efficiency by reducing public spending, cutting state subsidies and rationalising bureaucracy

sustainability – the desire by planners and agencies to avoid creating projects which depend on their continued support for success; also used in its environmental sense to ensure renewal of natural resources

targeting – the attempt to ensure that the benefits of a project reach a particular section of the population – women, farmers with no land, squatters, etc.

Third World – those countries that were seen as neither part of the capitalist First World nor the communist Second World in the period after the Second World War

top-down – interventions imposed on local people by those in authority – the opposite of bottom-up

trickle down – the assumption, which comes from neo-classical economics, that if economic growth is achieved then benefits will eventually 'trickle down' from the 'wealth producers' to the poorer sections of the population

Anthropological jargon

acculturation – originally used to refer to changes in cultures as they came into contact with each other, the term later became synonymous among US anthropologists with the idea that non-Western or 'indigenous' cultures went into decline after contact with industrialised ones

applied anthropology – the application of anthropological knowledge and research methodologies to practical issues, born out of anthropologists' involvement in colonial administration and development policy in the 1930s and 1940s

cultural relativism – derived from the work of Franz Boas (1858–1942), this concept encouraged anthropologists to understand each culture on its own terms, instead of making evolutionary or ethnocentric generalisations

diffusionism – a term associated with E.B. Tylor (1832–1917), used to explain the transmission of cultural traits across space, through culture contact or migration

discourse – based on the ideas of Michel Foucault (1926–84), discourse theory refers to the idea that the terms in which we speak, write and think about the world are a reflection of wider relations of power and, since they are also linked to practice, are themselves important in maintaining that power structure

ethnocentricity – the idea that a tendency exists to interpret other cultures according to the values of one's own, a term first used by William Sumner (1840–1910)

ethnography – a term which means both the study of a community or ethnic group at close quarters and the text (usually known as a monograph) which results

evolutionism – in contrast to diffusionists (see above), evolutionists believe that universal human psychological characteristics eventually produce similar cultural traits all over the world, although these evolve at different rates in different places

functionalism – a theory which tries to explain social and cultural institutions and relations in terms of the functions they perform within the system; heavily criticised because it fails to take account of historical factors such as change, conflict and disintegration

indigenous – used instead of the more pejorative 'native' to refer to the original inhabitants of an area which has been occupied by migrants; but still brings problems in many situations by implying that there are somehow 'legitimate' inhabitants of land with greater rights than newcomers

managerialism – the unwarranted and often ideological reliance on purely technical (as opposed to political) forms of problem solving

moral economy – a term associated with historian E.P. Thompson (1924–93) and used originally to refer to the idea that peasant political action was guided not just by economic hardship but sets of social, political and moral values and norms – subsequently the concept has been adapted and used beyond peasant studies

participant observation – the foundation of anthropological field research since the pioneering work of Malinowski (1884–1942), in which the anthropologist seeks to immerse herself as fully and as unobtrusively as possible in the life of a community under study

postmodernism – the wider cultural and epistemological rejection of modernity in favour of a broader pluri-cultural range of styles, techniques and voices, including the rejection of unitary theories of progress and scientific rationality. In anthropology in particular, postmodernism has led to the questioning of the authority of the ethnographic text and in part to a crisis of representation

structural-functionalism – a theoretical perspective associated with the British anthropologist Radcliffe-Brown (1881–1955), which stressed the importance of social relations and institutions in forming the framework of society, while at the same time functioning to preserve society as a stable whole

structuralism – following from the work in linguistics of Saussure and Jakobson, the anthropologist Lévi-Strauss (1908–2009) argued that that culture is a superficial manifestation of deeper structural principles, based on the universal human imperative to classify experience and phenomena

Acronyms

AAA – American Anthropology Association

BRAC – formerly the Bangladesh Rural Advancement Committee, now just used as an acronym

BRICs – Brazil, Russia, India and China

CSR – corporate social responsibility

DFID – Department for International Development (UK)

ECLA – Economic Commission of Latin America

FAO – Food and Agricultural Organisation

FSR – farming systems research

GAD – gender and development

IBRD – International Bank for Reconstruction and Development

IFAD – International Fund for Agricultural Development

ILO – International Labour Organization

IMF – International Monetary Fund

IPCC – Intergovernmental Panel on Climate Change

ITDG – Intermediate Technology and Development Group

MDGs – Millennium Development Goals

NGO – non-governmental organisation

ODA – Overseas Development Administration

OECD – Organisation for Economic Co-operation and Development

PLA – participatory learning and action

PRA – participatory rural appraisal

PRSP – poverty reduction strategy paper

SAPs – structural adjustment policies

SDA – social development advisers

SDGs – sustainable development goals

SIDA – Swedish International Development Cooperation Agency

SWAP – sector-wide approach

UNDP – United Nations Development Programme

UNICEF – United Nations Children's Fund

USAID – United States Agency for International Development

WID – women in development

WTO – World Trade Organization

Prelude: Development, Post-Development and More Development?

The idea of development stands like a ruin on the intellectual landscape. Delusion and disappointment, failures and crimes have been the steady companions of development and they tell a common story: it did not work. (Sachs, 1992: 1)

Welcome to 'Development World'. The pursuit of development has become a global concern and no one is unaffected. Aspiring to manage change in economic, political, social and cultural arenas, development is a world-shaping project. (Axelby and Crewe, 2013: ix)

Whatever happened to the anthropology of development and its 'postmodern challenge'? When we published the first edition of this book in 1996 (entitled *Anthropology, Development and the Post-Modern Challenge*) development was under sustained theoretical fire. Discredited for its evolutionary and Euro-American-centrism, and attacked by writers such as Wolfgang Sachs (1992) and Arturo Escobar (1995) for its role in the maintenance of postcolonial power relations, it seemed possible that in the next ten or twenty years development might expire altogether and that new framings of progressive change might arise. Even if reports of its death were greatly exaggerated, the era of 'post-development' thinking seemed to be upon us. Within anthropology postmodern critiques were also causing significant disquiet. Accused of creating exoticised representations of 'the other' and methodological techniques in which anthropologists subjugated and objectified the people they researched, the discipline seemed, for a while, in danger of losing its confidence, or even turning into a sub-field of literary critique, a direction suggested by Clifford and Marcus's 1986 book *Writing Culture*.

Anthropology, Development and the Post-Modern Challenge addressed these questions by arguing that the discipline should not balk at becoming directly involved with social problems. The book was a rallying cry for anthropological engagement in development, in all its varied meanings. In it, we argued that while the 'post-modern' attack on development was theoretically beguiling, it was in danger of contributing to an apolitical disengagement by anthropologists not wishing to dirty their hands with the dubious business of trying to change the world for the better. Indeed, we argued that Escobar's analysis homogenised and simplified development, which by the 1990s involved a lot more than colonial-style planners pushing people around. Instead, we suggested that while they had previously been treated within the discipline as working in an inferior sub-field, anthropologists *of* and *in* development had much to offer. What they had to offer was not only useful to what Axelby and Crewe (2013) call 'Development World', for example in helping to formulate new policies and practices which prioritised issues of power, poverty and inequality rather than economic growth or modernisation, but these anthropologists also had an important role to play in bringing rich insights around the relationship between social relations and economic change to academic anthropology itself.

Tracing the links between ethnographic work and new, often (at the time) radical directions in development we argued that anthropology's influence in shaping new formulations that moved far beyond the monolithic colonial discourse described by Sachs, Escobar and others was potentially huge. The questions of access, control and effects that anthropologists asked informed development work that had power, unequal access and inequality as its focus. Ethnographic methods were key to new ways of seeing and doing. Indeed, based on anthropological practice, new techniques for gathering information and using it to effect change were fast catching on in Development World, such as the various participatory learning and action (PLA) approaches that became popular during the 1990s.

So, what happened? Was Escobar right in his prediction that development was reaching its end game? Over the last twenty years the rate and substance of change is remarkable in two ways. In the first sense, it has been profound and rapid. The world we described in 1996 was very different from the contemporary post-9/11 era of war, securitisation and, more recently, financial meltdown, recession and austerity. While the so-called 'BRICs' (Brazil, Russia, India, China) have been hailed as emerging economies for their rapid rates of economic growth, industriali-sation and urbanisation, other countries, most notably in southern Europe, have experienced dramatic de-development as a result of the global financial crisis. Where and with whom development work is supposed to take place is increasingly blurred, as distinctions between 'the developed' and 'undeveloped' world become increasingly problematic. Meanwhile not only have new governments, agencies and donors entered the fray as givers of aid and do-ers of development – India and China are obvious examples – but ethical conduct and schemes of improvement have been taken on by corporations as a badge of honour. Great moral value is placed on improving the lives of others, especially if they live on the other side of the world. Today, philanthropy is the hobby of choice for billionaires, pop stars and actors, who rush to endorse projects and causes while making 'poverty history' or 'turning oppression into opportunity for women worldwide'[1] at the click of a mouse, paying a donation or wearing a wristband.

In the second sense, change is remarkable only for its absence. As Axelby and Crewe (2013) insist, development – both as a concept and a set of practices – still continues to wield huge power globally. To this extent the story remains the same, regardless of the stones thrown from both sides of the political divide. Development's capacity to absorb critiques and turn what were once radical alternatives into de-politicised common practices while on the central stage business goes on as normal is testimony to its enduring power. This applies equally to its ideological dominance: the view that economic growth and other measures will lead to the social good, generally imagined as involving Westernisation or modernisation. The United Nations Millennium Goals (MDGs), for example, are, naturally, *development* goals, as are the post-2015 Sustainable Development Goals (SDGs) that have been designed to follow on from these.[2] Meanwhile the development industry continues unabated. Universities continue to offer degrees in development studies, government Departments of Development remain in place, civil society and non-governmental organisations (NGOs) are tasked to carry out development, and many thousands of experts, consultants, fieldworkers and officials rely upon it for their livelihood. And it continues to touch the lives of almost everyone.

So what of the anthropology of and in development? In this second edition of *Anthropology, Development and the Post-Modern Challenge* we argue that anthropology's potential to analyse and describe processes of change, and contribute to alternative visions, remains as powerful as ever. As we shall see in the chapters that follow, rather than being on the margins, the anthropology of development has in many ways been increasingly absorbed into the mainstream. In the twenty years since the book was first published the field has become enormous. This is partly because so many anthropologists are working in contexts of rapid economic transformation, globalisation and cultural complexity that questions of change are impossible to ignore or corral as a sub-field. Today only a small minority (indeed, a handful?) of anthropologists engage in research with communities or groups that could be considered to be unaffected by the wider world. Even if this were not the case, what matters more than the context in which fieldwork is carried out are the questions that anthropologists ask. That the world is changing is nothing new. What *has* shifted is not only the rate and scale of change, but anthropology's willingness to acknowledge, document and theorise it. Anthropological studies are nowadays framed by questions concerning global capitalism, conflict, governance, migration and environment, to name a few of the most obvious. If the anthropology *of* development is in its broadest terms the anthropology of change and transition, or of global economic systems, then we are hard pressed to identify where the boundaries lie between this and the anthropological mainstream.

There remains, however, a body of work that is clearly *of* development. Here we need to pay attention to human geographer Gillian Hart's (2001: 650) helpful distinction:

> between 'big D' Development defined as a post-second world war project of intervention in the 'third world' that emerged in the context of decolonization and the cold war, and 'little d' development or the development of capitalism as a geographically uneven, profoundly contradictory set of historical processes.

One meaning refers to development as unfolding capitalist change and the other to the intentional, planned change that takes place within 'Development World' (Axelby and Crewe, 2013). We shall be using this distinction between big D and little d throughout the book.[3] We suggest that early twenty-first-century anthropology of development is now animated by questions that pick up where Escobar et al. left off to study post-development (see, for example, Escobar, 2008, 2010). This newly animated anthropology of development is a large and rapidly expanding field of study of which we can only describe a fraction of the work being undertaken. What does it mean to analyse development work and knowledge as a discursive field? How might we understand policy and projects, not to say the cultural worlds of those who produce them, in these terms? Centrally, how might we understand new approaches to development, including micro-finance, Fair Trade and entrepreneurial schemes that aim to mine the market potential of the 'bottom of the pyramid' – in Prahalad's (2004) influential phrase – as linked to global governance and the ongoing hegemonic might of global capitalism? How might we understand the developmentalisation of welfare – now increasingly relabelled wellbeing – in the form of large-scale investments in social protection interventions, such as conditional cash transfer schemes, micro-insurance and commodity subsidies? What moral economies underlie these schemes, and how can anthropological theory be used to explain them?

For both development and Development, we argue that the questions of *access*, *control* and *effects* that we outlined in the first edition remain core concerns. Since these primary questions are the same, we have left some of the original book intact: who gains and who loses? What is happening or has happened and why? What are the underlying dynamics of power at all levels and scales? As before, we argue that these seemingly simple questions must remain at the heart of anthropological analysis of both senses of d/Development, to be used as an ethical yardstick both for those working within the industry, and for those using academic research to critique it. Indeed, while once derided as the discipline's 'evil twin' (Ferguson, 1997) the anthropology of D/development has more recently been celebrated as its 'moral narrative' (Gow, 2002). In a world increasingly divided between haves and have-nots, where profit and economic growth seem invariably to involve disenfranchisement and exclusion for those at the margins, and where morality and personhood are played out via consumption it is anthropologists who are best placed to offer empirical evidence and analysis of how global systems work and what this means for ordinary people. It is also anthropologists, armed with cross-cultural perspectives, who are able to offer fresh ways of seeing, to combat the accepted orthodoxies: for example that 'the market' can cure the world's ills or that global capitalism's economic systems are rational (see Graeber, 2011).

Here it is useful to distinguish between anthropology as critique ('of') and anthropology as enabling or involving action, or 'in' development (Hale, 2006). As with the first edition of the book, we have divided the chapters with this distinction in mind, offering an updated account of the history of engaged anthropology in general and anthropologists 'in' development in particular, along with an overview of academic analyses and ethnographies 'of', adding new sections to cover the directions the field has taken since the late 1990s. But again, while useful for thinking about different types of activity – the first using anthropological perspectives in order to critique, the second using anthropological perspectives in order to change things– the boundaries between 'of' and 'in', or 'pure' and 'applied' are ever more fuzzy. This is partly because in recent years the political pressures to justify academic research have been piling up, with an increasingly instrumentalist approach being taken by funders and policy-makers. In Britain, for example, bureaucratic exigencies for academics to produce 'impact' in order to gain funding and as part of government audits of 'academic excellence' have forced anthropologists to think more seriously about the effects that their work has beyond academia.

The problem with the separation between 'pure' and 'applied' is the implication that research 'for research's sake' is somehow locked in a box marked 'ivory tower' and has no value beyond that. Yet, as we argued in the first edition of this book, ideas matter. Merely because the person producing research is not the same person who is designing policy, protesting or leading social movements does not mean that the research has not played an influential role in the actions that ensue. While there are clearly anthropologists who are research-oriented and those who are actively engaged in policy, advocacy or protest, 'pure' research leads to insights and ideas that go on to inform 'applied' work. If in any doubt of how scholarly critique and theory can lead to change, one need only think of the ideas of Karl Marx!

What form has this synergy between research, ideas and action taken in recent years? Writing in the mid-1990s we were excited by the possibilities that anthropology offered in changing development agendas. Ideas from feminist anthropology, research into the interface between scientific and indigenous knowledge and the ethnographic focus on everyday lives and perspectives, not to say power dynamics, contributed to an array of new directions that offered the hope of breaking free from the dominant discourse. Since then there have been successes but also failures. Many of the most radical ideas – empowerment and participation, for example – have been taken on, absorbed but also de-politicised by development. This can be read as a sign of positive change, but in many instances has involved a watered down version of the original ideas. In some cases potentially radical practices have been reduced to 'box-ticking' exercises. In this sense, Escobar was right in his prediction that the discourse would always absorb change, yet remain essentially the same, writing that 'the new discourses exist in the same plane of the original concept, and this contributes to the discourse's self-creation and autoreferentiality' (1995: 201). Yet we end by outlining new ideas and new approaches which offer alternatives as well as critiques of mainstream development, and which meet the challenges of the twenty-first century.

The book is structured as follows. After our introductory Chapter 1, which provides a brief outline of development theory, and the changes to Development World that have taken place since 1996, we move in Chapter 2 to an updated account of the histories of applied and 'engaged' anthropology, as well as its more recent appearances in the shape of 'protest' and/or 'anarchic' anthropology. This spans anthropological involvement with development but also its wider role in advocacy, protest and action. In Chapter 3, we provide a review of the anthropology 'of' d/ Development, both up to 1996 when *Anthropology, Development and the Post-Modern Challenge* was published, and in the years that have followed. After having scoped out and updated the field of study and action, Chapter 4 moves to the ideas and approaches which, back in the mid-1990s, we argued provided ways of breaking out of the discursive strictures of mainstream development. Since it is our argument that the core questions of access, control and effects remain the same, we have left this chapter mostly in its original version. In Chapter 5 we revisit some of the ideas that, in the 1990s, we argued held the potential for progressive change within development, showing how in some quarters these have been reduced to what Cornwall and Eade (2010) have termed 'buzzwords and fuzzwords'. Despite these problems, we argue in the Conclusion that this does not mean that, in other contexts, the potential of these ideas, or for other forms of anthropological engagement, is any less. Rather, anthropology can critique more conventional projects and transcend the confines of projects carried out in the global South. It has much to say also about the global North, about new forms of development and about transcending development altogether to provide fresh insights concerning the nature of empowerment, participation and gender.

In writing this new edition, our aim is also to provide the reader with a sense of the historical depth of the anthropology of, and in, D/development. For this reason we have included some of the original sections of the 1996 edition while adding up-to-date accounts of what has happened since. As before, we have been selective in our coverage of the wide-ranging field of development and, in general, have adopted the policy of sticking to what we know. As a result, we again do not engage in any depth with issues such as medical anthropology, environmental sustainability or climate change, or with the worlds of international humanitarian assistance. The world has changed a great deal but the core questions that anthropologists ask remain largely the same. The need for an engaged anthropology is as pressing as ever.

1

Understanding Development: Theory and Practice into the Twenty-First Century

What is 'development'? How did it become both so important and so problematic? In this introductory chapter we outline the history of development and its theoretical underpinnings, from its Enlightenment origins to the present before asking what became of 'the postmodern challenge'? The chapter is broadly presented in two parts. The first half of the chapter spans the period up to 1996 when the first edition of this book was published. It introduces the 'aid industry', analyses the history of the idea of development, and discusses the rise and fall of its grand theories. In the second part, we discuss some of the wide-ranging changes that have taken place during the past two decades, both within the world at large, where the balance of global power has shifted in significant ways, inequalities both within and between nations have increased, and where there has been a post 9/11 policy emphasis on securitisation; and also within the world of development. This has seen an increased and growing role for the private sector, an increased emphasis on managerialism and results in the development intervention field, and the rise of non-Western donor countries such as China that offer low-income countries new choices in relation to aid and projects.

Development: history and meanings

In virtually all its usages, development implies positive change or progress. It also evokes natural metaphors of organic growth and evolution. The *Oxford Dictionary of Current English* defines it as 'stage of growth or advancement' (1988: 200). As a verb it refers to activities required to bring these changes about, while as an adjective it is inherently judgemental, for it involves a standard against which things are compared. While 'they' are undeveloped, or in the process of being developed, we (it is implied) have already reached that coveted state. When the term was used by President Truman in a speech in 1949, vast areas of the world were therefore suddenly labelled 'underdeveloped' (Esteva, 1993: 7). A new problem was created, and with it the solutions; all of which depended upon the rational-scientific knowledge of the so-called developed powers (Hobart, 1993: 2).

Capitalism and colonialism: 1700–1949

The notion of development goes back further than 1949, however. Larrain has argued that while there has always been economic and social change throughout history, consciousness of 'progress', and the belief that this should be promoted, arose only within specific historical circumstances in northern Europe. Such ideas were first generated during what he terms the 'age of competitive capitalism' (1700–1860): an era of radical social and political struggles in which feudalism was increasingly undermined (Larrain, 1989: 1).

Concurrent with the profound economic and political changes that characterised these years was the emergence of what is often referred to as the 'Enlightenment'. This social and cultural movement, which was arguably to dominate Western thought until the late twentieth century, stressed tolerance, reason and common sense. These sentiments were accompanied by the rise of technology and science, which were heralded as ushering in a new age of rationality and enlightenment for humankind, as opposed to what were now increasingly viewed as the superstitious and ignorant 'Dark Ages'. Rational knowledge, based on empirical information, was deemed to be the way forward (Jordanova, 1980: 45). During this era polarities between 'primitive' and 'civilised', 'backward' and 'advanced', 'superstitious' and 'scientific', 'nature' and 'culture' became commonplace (Bloch and Bloch, 1980: 27). Such dichotomies have their contemporary equivalents in notions of undeveloped and developed.

Larrain links particular types of development theory with different phases in capitalism. While the period 1700–1860 was characterised by the classical political economy of Smith and Ricardo and the historical materialism of Marx and Engels, the age of imperialism (1860–1945) spawned neo-classical political economy and classical theories of imperialism. Meanwhile, the subsequent expansionary age of late capitalism (1945–66) was marked by theories of modernisation, and by the crises of 1966–80 by neo-Marxist theories of unequal exchange and dependency (Larrain, 1989: 4). We shall elaborate on these later theories further on in this chapter.

While capitalist expansion and crisis are clearly crucial to the history of development theory, the latter is also related to rapid leaps in scientific knowledge and social theory over the nineteenth and early twentieth centuries. A key moment in this was the publication of Charles Darwin's *Origin of Species* in 1859. This was to have a huge influence on the social and political sciences in the West. Inspired by Darwin's arguments about the evolution of biological species, many political economists now theorised social change in similar terms. In *The Division of Labour* (1947, originally published in 1893), for instance, Emile Durkheim – who is now widely considered one of the founding fathers of sociology – compared 'primitive' and 'modern' society, basing his models on organic analogies. The former, he suggested, is characterised by 'mechanical solidarity', in which there is a low division of labour, a segmentary structure and strong collective consciousness. In contrast, modern societies exhibit 'organic solidarity'. This involves a greater interdependence between component parts and a highly specialised division of labour: production involves many different tasks, performed by different people; social structure is differentiated, and there is a high level of individual consciousness.

Although their work was quite different from Durkheim's, Marx and Engels also acknowledged a debt to Darwin (Giddens, 1971: 66). Marx argued that societies were transformed through changes in the mode of production. This was assumed to evolve in a series of stages, or modes of production, which Marx believed all societies would eventually pass through. Nineteenth-century Britain, for example, had already experienced the transformation from a feudal to a capitalist mode of production. When capitalism was sufficiently developed, Marx argued, the system would break down and the next stage – of socialism – would be reached. We shall discuss below the influence of Marxism on theories of development.

Closely associated with the history of capitalism is of course that of colonialism. Particularly over later colonial periods (say, 1850–1950), notions of progress and enlightenment were key to colonial discourses, where the 'natives' were constructed as backward or childlike, and the colonisers as rational agents of progress (Said, 1978: 40). Thus while economic gain was the driving force for imperial conquest, colonial rule in the nineteenth and twentieth centuries also involved attempts to change local society with the introduction of European-style education, Christianity, and new political and bureaucratic systems. Notions of moral duty were central to this, often expressed in terms of the relationship between a trustee and a minor (Mair, 1984: 2). While rarely phrased in such racist terms, development discourse in the 1990s often involves similar themes: 'good government', institution building and gender training are just three currently fashionable concerns which promote 'desirable' social and political change. With these concepts arising from such dubious beginnings, it is hardly surprising that many people today regard them with suspicion.

By the early twentieth century the relationship between colonial practice, planned change and welfarism had become more direct. In 1939, for example, the British government changed its Law of Development of the Colonies to the Law of Development and Welfare of the Colonies, insisting that the colonial power should maintain a minimum level of health, education and nutrition for its subjects. Colonial authorities were now to be responsible for the economic development of a conquered territory, as well as the wellbeing of its inhabitants (Esteva, 1993: 10).

The postcolonial era: 1949 onwards

Notions of development are clearly linked to the history of capitalism, colonialism and the emergence of particular European epistemologies from the eighteenth century onwards. In the latter part of the twentieth century, however, the term has taken on a range of specific, although often contested, meanings. Arturo Escobar (1988, 1995) argues that it has become a discourse: a particular mode of thinking, and a source of practice designed to instil in 'underdeveloped' countries the desire to strive towards industrial and economic growth. It has also become professionalised, with a range of concepts, categories and techniques through which the generation and diffusion of particular forms of knowledge are organised, managed and controlled (Escobar, 1995). We shall be returning to Escobar's views of development as a form of discourse, and thus of power, later on in this book. For now, let us examine what these more contemporary post-Second World War meanings of development involved.

When President Truman referred in 1949 to his 'bold new programme for making the benefits of our scientific advances and industrial progress available for the improvement and growth of underdeveloped areas' (cited in Esteva, 1993: 6), he was keen to distance his project from old-style imperialism. Instead, this new project was located in terms of economic growth and modernity. Countries, it was assumed, would be able to move through a set of stages of economic growth during which the constraints of traditional forms of social organisation would give way to the modern values of individualism and innovation that would allow capitalist economic development to flourish. During one of the earliest missions of the newly formed International Bank for Reconstruction and Development (IBRD) to Colombia in 1949, for example, it was integrated strategies to improve and reform the economy that were called for, rather than the consideration of social or political changes. The definition of development as economic growth has remained central to mainstream thinking ever since, through the debt crises of the 1980s and the subsequent imposition on poor countries of 'structural adjustment' programmes by the World Bank and International Monetary Fund (IMF), and the 1990s good governance policy agendas. Despite the challenge from some sections of the UN in the form of the idea of 'human development' that was put forward by Mahbub ul Haq and Amartya Sen, and the rise of 'poverty reduction' as a mainstream development objective during this period, the idea of development as economic growth never really went away. It returned to high prominence once again in the 2000s in studies such as Dollar and Kraay's (2002) influential study *Growth Is Good for the Poor*, which became one of the World Bank's most widely cited publications (de Haan, 2009).

Behind these aims is the assumption that growth involves technological sophistication, urbanisation, high levels of consumption and a range of social and cultural changes. For many governments and experts the route to this state was industrialisation. As we shall shortly see, this is closely linked to theories of modernisation. Successful development is measured by economic indices such as the Gross National Product (GNP) or per capita income. It is usually assumed that this will automatically lead to positive changes in other indices, such as rates of infant mortality, illiteracy, malnourishment and so on. Even if not everyone benefits directly from growth, the 'trickle-down effect' will ensure that the riches of those at the top of the economic scale will eventually benefit the rest of society through increased production and thus employment. In this understanding of development, if people become better fed, better educated, better housed and healthier, this is the indirect result of policies aimed at stimulating higher rates of productivity and consumption, rather than of policies directly tackling the problems of poverty. Development is quantifiable, and reducible to economics. One major drawback to defining development as economic growth is that in reality the 'trickle-down effect' rarely takes place; growth does not necessarily lead to enhanced standards of living. As societies in the affluent North demonstrate, the increased use of highly sophisticated technology or a fast-growing GNP does not necessarily eradicate poverty, illiteracy or homelessness, although it may well alter the ways these ills are experienced. In contrast, neo-Marxist theory, which was increasingly to dominate academic debates surrounding development in the 1970s, understands capitalism as inherently inegalitarian. Economic growth thus by definition means that some parts of the world, and some social groups, are actively underdeveloped. Viewed in these terms, development is an essentially political process; when we talk of 'underdevelopment' we are referring to unequal global power relations.

Although the modernisation paradigm continued to dominate mainstream thought, this definition of development – as resulting from macro and micro inequality – was increasingly promoted during the 1970s and, within some quarters, throughout the 1980s. It can be linked to what became termed the 'basic needs' movement, which stressed the importance of combating poverty rather than promoting industrialisation and modernisation. Development work, it was argued, should aim first and foremost at satisfying people's basic needs; it should be poverty-focused. For some, this did not involve challenging wider notions of the ultimate importance of economic growth, but instead involved an amended agenda in which vulnerable groups such as 'small farmers' or 'women-headed households' were targeted for aid. Many of these projects were strongly welfare-oriented and did not challenge existing political structures (Mosley, 1987: 29–31).

In the 1990s the desirability of technological progress was further questioned. Environmental destruction in the form of deforestation, pollution and ozone depletion was an increasingly pressing issue and sustainability became an important development buzzword. By the 2000s climate change was also becoming a development issue, as studies by the Intergovernmental Panel on Climate Change (IPCC) began to receive more attention (see Castro et al., 2012). Cases where technological change has been matched by growing inequality and the breakdown of traditional networks of support are now so well documented as to be standard reading on most undergraduate courses on development. It is becoming clear that mechanisation and industrialisation are mixed blessings, to say the least. Combined with this, the optimism of the 1960s and early 1970s, when many newly independent states were striving for rapid economic growth, was replaced by increasing pessimism during the 1980s. Faced by debt, the inequality of international trading relations and in many cases political insecurity, many governments, particularly those in Africa and Latin America were forced to accept the rigorous structural adjustment programmes insisted upon by the World Bank and IMF.

Development involved the construction not only of particular ideas, but also of a set of specific practices and institutions. Before turning to the various theories that have been offered since 1949 to explain development and underdevelopment, let us therefore briefly turn to what is often referred to as 'the aid industry'.

The 'aid industry'

As we have already indicated, aid from the North to the South was without doubt a continuation of colonial relations, rather than a radical break from them (Mosley, 1987: 21). Donors today tend to give most aid to countries which they previously colonised: British aid is concentrated mostly upon South Asia and Africa, while the Dutch are heavily involved in South East Asia, for example. Although planning is a basic human activity, the roots of planned development were planted during colonial times, through the establishment of bodies such as the Empire Marketing Board in 1926 and the setting up of Development Boards in colonies such as Uganda (Robertson, 1984: 16). The concept of aid transfers being made for the sake of development first appeared in the 1930s, however. Notions of mutual benefit, still prevalent today, were key. The aim was primarily to stimulate markets in the colonies, thus boosting the economy at home (Mosley, 1987: 21).

Despite these initial beginnings, the real start of the main processes of aid transfer is usually taken to be the end of the Second World War, when the major multilateral agencies were established. The IMF and the IBRD (later to become the World Bank) were set up during the Bretton Woods Conference in 1944, while the Food and Agricultural Organization (FAO) was created as a branch of the United Nations in 1945. In contrast to what became known as 'bilateral aid', which was a transfer from one government to another, 'multilateral aid' came to involve a number of different donors acting in combination, none of whom (supposedly) directly controls policy. However, from the outset donors such as the World Bank were heavily influenced by the US and tended to encourage centralised, democratic governments with a strong bias towards the free market (Robertson, 1984: 23). Meanwhile, various bilateral agencies were also established by the wealthier nations. These are the governmental organisations, such as the United States Agency for International Development (USAID; set up in 1961) or the British Department for International Development (DFID; previously the Overseas Development Agency [ODA], established as the Overseas Development Ministry in 1964), both of which are involved in project and programme aid with partner countries. Figure 1.1 shows the different main organisational actor types in the aid system.

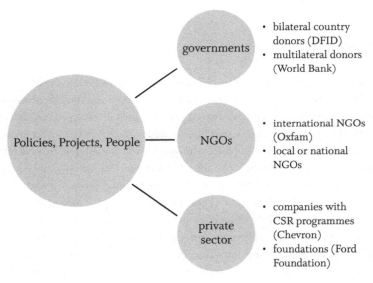

Figure 1.1   Agencies in the aid system

Considerable amounts of aid were initially directed at areas in Europe which were devastated after the Second World War. By the early 1950s the Cold War made aid politically attractive for governments anxious to stem the flow of communism in the South. During this period the World Bank changed its focus from reconstruction to development. By the late 1960s, after many previously French and British colonies had gained independence, aid programmes expanded rapidly. Indeed, rich donor countries actually began to come into competition with each other in their efforts to provide assistance to poor countries, a clear sign of the economic and political benefits which accompanied aid. Keen to improve their product, many now stressed development, instigating grandiose and prestigious schemes. The 1960s also saw the first UN Decade for Development, with a stated aim of achieving 5 per cent growth rates, and 0.7 per cent of donor countries' GNP being given in aid. In 1984/5 the US gave 0.24 per cent, the UK 0.34 per cent, and Norway 1.04 per cent (Cassen et al., 1986: 8). Today few countries give this much, at least in per capita terms, with Scandinavian countries the most generous. In the UK, successive governments in recent years have aimed to raise aid spending in order to meet the 0.7 target which was finally achieved in 2014, though by this time many were questioning the relevance of such targets.[1]

Since the earliest days of the aid industry, there have been significant shifts in those countries giving and receiving the most aid. Sub-Saharan Africa receives the largest proportion of aid, whereas earlier it was India that was the largest recipient. Likewise, more and more countries have been so successful that they are now becoming influential donors: Japan and Saudi Arabia are examples. We explore these changes later in this chapter.

Whether or not aid is a form of 'neo-imperialism' has been a moot point in development studies. Some writers argue that aid is simply another way in which the political and economic power of the North continues to be asserted over the South, developing only the dependency of recipients on their donors (for example, Hayter, 1971; Sobhan, 1989). More recent aid critics such as William Easterly (2006) and Dambisa Moyo (2009) argue that aid distorts markets, sustains corruption and remains unhelpfully top-down. Others stress that, while there are undoubted benefits to donors (political influence perhaps, or the creation of markets for domestically produced products), aid cannot simply be understood as exploitative. Most aid, for example, is aimed at the neediest countries, rather than the biggest potential markets and allies, and many projects and programmes are planned with good intentions and genuine aims to promote desirable change (Mosley, 1987). Indeed, rather than the wholly negative picture presented by polemicists such as Hancock in his attack on the aid industry (1989), some writers have argued that most aid is successful in terms of its own objectives (Cassen et al., 1986). Others maintain a middle line, pointing out the complex reasons why aid projects fail and constructively suggesting how they could help, rather than accusing them all of being neo-imperial façades, and thus all 'bad' (Mosley, 1987; Madeley, 1991; Riddell, 2008).

An interesting twist to these debates was given by Ferguson (1990) in his account of the development regime in Lesotho, part of which we discuss in Chapter 3. Ferguson argues that, rather than deliberately setting out to perpetuate neo-colonial relationships between the North and South (for example, by bringing peasants into the global market under unfavourable terms of exchange, as some political economists have argued, or by securing markets for goods produced in the donor country), the role of aid projects is actually far more subtle:

> Whatever interests may be at work, and whatever they may think they are doing, they can only operate through a complex set of social and cultural structures so deeply embedded and so ill-perceived that the outcome may be only a baroque and unrecognisable transformation of the original intention. The approach adopted here treats such an outcome as neither an inexplicable mistake, nor the trace of a yet-undiscovered intention, but as a riddle, a problem to be solved, an anthropological puzzle. (Ferguson, 1990: 17)

Ferguson's contribution is therefore to distinguish between the *intentions* of those working in the aid industry and the *effects* of their work. As such it provides a very useful way of moving beyond the simple rhetoric of the 'aid as imperialism' school of thought.

Following on from Ferguson's approach, we do not think it worthwhile to spend too much time considering whether aid is or is not a 'good' thing. Instead, we assume that it exists and will continue to exist for some time. Rather than simply condemning aid and development work, what we are concerned with is how anthropology might be used to critique, improve and suggest alternatives to it. How this might be done is a central theme of this book. Before exploring these issues further, let us turn to a brief summary of the different theoretical perspectives informing developmental work.

Theories of development

Conventionally, development theory is described in terms of two oppositional paradigms, both of which involve a range of different measures. Like most 'grand theories', neither has stood up well to the onslaught of 1990s postmodernism. Today, there is no single theoretical model which is commonly used to explain development, nor is there any one 'solution' to the problems of underdevelopment. Indeed, contemporary understandings tend to draw from a variety of theoretical sources and suggest a variety of strategies.

Modernisation

What can be labelled 'modernisation theory' is a collection of perspectives which, while at their most intellectually influential in the 1950s and 1960s, continue to dominate development practice today. Many of the technicians and administrators involved in project planning are still essentially modernisers, even if their jargon is more sophisticated than that of their predecessors in the 1960s. As Norman Long and Ann Long (1992: 18) have put it modernisation 'visualises development in terms of a progressive movement towards technologically more complex and integrated forms of "modern" society'. Likewise, many development economists still pin their hopes to the promises of modernisation. For example, one of the most influential recent books on development, *Why Nations Fail* by Daron Acemoglu and James Robinson (2012), offers a powerful challenge to mainstream development policy by arguing that it fails to recognise sufficiently the constraints placed on policy-making by predatory or 'extractive' political institutions; the authors argue instead for the importance of building 'inclusive' institutions, such as formal property and rights systems and effective court systems – but the argument remains one that is directed towards economic growth as the key to development as 'prosperity'.

Industrialisation, the transition from subsistence agriculture to cash-cropping, and urbanisation are all keys to this process. Modernisation is essentially evolutionary; countries are envisaged as being at different stages on a linear path which leads ultimately to an industrialised, urban and ordered society. Much emphasis is put upon rationality, in both its economic and moral senses. While modern, developed societies are seen as secular, universalistic and profit-motivated, undeveloped societies are understood as steeped in tradition, particularistic and unmotivated to profit, a view exemplified by George Foster's (1962) work on the idea of the 'peasant's image of the limited good'.

As we have already seen, these ideas have roots in nineteenth- and early twentieth-century political economy, much of which sought to theorise the sweeping social and economic changes associated with industrialisation.

Durkheim's model of an industrialised 'organic' society, Simmel's thoughts on the increasingly pervasive monetary economy and Weber's discussion of the relationship between Protestantism and industrial capitalism are all examples. Later during the 1960s the work of economist W.W. Rostow illustrated the concept of modernisation par excellence. In his works on economic growth (Rostow, 1960a, 1960b), the forms of growth already experienced in the North are taken as a model for the rest of the world. While economies are situated at different stages of development, all are assumed to be moving in the same direction. Traditional society is poor, irrational and rural. The 'take-off' stage requires a leap forward, based on technology and high levels of investment; preconditions for this are the development of infrastructure, manufacturing and effective government. After this point, societies were believed to reach a stage of 'self-sustaining' growth; in its 'mature' stage, technology pervades the whole economy, leading to 'the age of high mass consumption', high productivity and high levels of urbanisation (Robertson, 1984: 25).

Some writers have attached particular social characteristics to the different stages, often with evolutionary overtones. For example, Talcott Parsons has argued that nuclear families are best suited to the highly mobile, industrialised world (Parsons, 1949). Others associate industrial society with (again) rational political systems, realism and the death of ideology (Kerr et al., 1973, cited in Robertson, 1984: 33). Interestingly, early feminist work on the relationship between capitalist growth and gender, while usually critical of development, also sometimes implied that stages in the development process were associated with particular forms of gender relations, most notably to do with changes in the division of labour (for example, Boserup, 1970; Sacks, 1975).

If one believes that life is generally better in the Northern countries than in their poorer neighbours in the South (which in terms of material standards of living cannot easily be denied), modernisation is an inherently optimistic concept, for it assumes that all countries will eventually experience economic growth. This optimism must be understood in the historical context of post-war prosperity and growth in the North, and independence for many Southern colonies in the 1950s and 1960s. The governments of many newly independent countries, like their ex-colonisers, often believed that – with a little help – development would come swiftly, and many launched ambitious five-year plans to this effect (for example, India's First Five-Year Plan in 1951, and Tanzania's First Five-Year Plan in 1964). Truman's speech embodies this initial optimism.

Another reason why modernisation can be described as optimistic is that it presents development as a relatively easy process. Enduring under-development is explained in terms of 'obstacles'. These are internal to the countries concerned, ideologically neutral, and can generally be dealt with pragmatically. Inadequate infrastructure is a good example. Factors conventionally used to explain this are lack of capital, weak or corrupt management and lack of local expertise (both of which might cause roads and bridges not to get built, or to be badly maintained) and, perhaps, difficult environmental conditions (mountainous terrain, continuous flooding). The solutions to these problems are straightforward: roads and bridges can be built with external capital and expertise in the form of aid donated by the developed North; local technicians and bureaucrats can be trained, and 'good government' supported. Another strategy to improve infrastructure might be the introduction of information technology to local institutions, or the training of personnel to use new technology. In both scenarios, various changes are understood as necessary for a country or region to 'take-off'. With more efficient infrastructure, economic growth is encouraged and, it is hoped that, barring other obstacles, the country will move on to the next stage. Development agencies and practitioners are thus cast in the role of trouble-shooters, creating a range of policies aimed at 'improvement' (Long, 1977).

By the late 1960s it was becoming obvious that, despite attempts to remove obstacles to development, often involving considerable foreign capital investment, economic growth rates in developing countries were disappointing; in some cases there were even signs that poverty was increasing. The failure of several large-scale development projects, which should have prompted 'take-off', increasingly indicated that simplistic notions of modernisation were inadequate. One now notorious case is the Groundnut Scheme of southern Tanzania. This latter project received £20 million in 1946–52 (the total British aid budget in 1946–56 was £120 million) and had a return of zero (Mosley, 1987: 22). Unquestioning faith in the desirability of cash crops on the part of planners, together with inadequate research into local farmers' needs and into the appro-priateness of different crops to the local environment, was central to the scheme's failure.

Modernisation, as both a theory and a set of strategies, is open to criticism on virtually every front. Its assumption that all change inevitably follows the Western model is both breathtakingly ethnocentric and empirically incorrect, a fact which anthropologists should have little difficulty in spotting. Indeed, anthropological research has continually shown that economic development comes in many shapes and forms; we cannot generalise about transitions from one 'type' of society to another. Religious revivalism is just one example of this, and has been interpreted as a reaction to modernity (see, for example, Ahmed, 1992). Combined with this, while theories of modernisation assume that local cultures and 'peasant' traditionalism are obstacles to development, what Norman Long calls 'actor-oriented research' (1977; see also Long and Long, 1992) has consistently found that, far from being 'irrational', people in poor countries are open to change if they perceive it to be in their interest. They often know far better than development planners how to strategise to get the best from difficult circumstances, yet modernisation strategies rarely, if ever, pay heed to local knowledge. Indeed, local culture is generally either ignored by planners or treated as a 'constraint'. This is a grave failing, for anthropologists such as Mair (1984) and Hill (1986) have shown in detail how an understanding of local culture is vital for more appropriate development projects. We shall spend much of this book discussing such insights.

Modernisation also ignores the political implications of growth on the micro level. Premised on the notion of 'trickle down', it assumes that once economic growth has been attained, the whole population will reap the rewards. Again, anthropologists and sociologists have repeatedly shown that life is not so simple. Even in regions of substantial economic growth, poverty levels often remain the same, or even increase further (Mosley, 1987: 155). Evidence from areas which have experienced the so-called Green Revolution illustrates how, even when many of the signs of economic development are present, localised poverty and inequality can persist (see Pearse, 1980). Disastrously (for the poorest or for some minorities), modernisation theory does not distinguish between different groups within societies, either because it assumes these to be homogeneous (the 'mass poor') or because it believes that eventually the benefits of growth are enjoyed by all. The communities which are at the receiving end of development plans are, however, composed of a mixture of people, each with different interests and levels of access to power and resources (Hill, 1986: 16–29). Heterogeneity exists not only between households but also within them. The marginalisation of women by development projects, which treat households as equal and homogeneous units, is a case in point (Whitehead, 1981; Rogers, 1980; Ostergaard, 1992).

The most fundamental criticism of theories of modernisation, however, is that they fail to understand the real causes of underdevelopment and poverty. By presenting all countries as being on the same linear path, they completely neglect historical and political factors which have made the playing field very far from level. Europe during the Industrial Revolution and Africa or South Asia in the second half of the twentieth century are not, therefore, comparable. These points have been forcibly made by what is generally referred to as dependency, or neo-Marxist, theory. This school of thought was radically to affect development studies during the 1970s.

Dependency theory

One of the first groups to explain development in terms of political and historical structures was the Economic Commission of Latin America (ECLA). Established in 1948 by the United Nations, by the 1950s this had become a group of radical scholars whose outlook was deeply influenced by Marxism. The work of the ECLA drew attention to the structure of under-development: unequal relations between the North and South, especially in terms of trade, the protectionism of many Northern economies and the dependency on export markets of many countries within Latin America. These notions of dependency and underdevelopment (as opposed to undevelopment) gained widespread recognition with the work of A.G. Frank (1967).

Drawing from Marxist concepts of capitalism as inherently exploitative, dependency theorists argue that development is an essentially unequalising process: while rich nations get richer, the rest inevitably get poorer. Like most Marxist analysis, their work is primarily historical and tends to focus upon the political structures that shape the world. Rather than being undeveloped, they argue, countries in the South have been underdeveloped by the processes of imperial and post-imperial exploitation. One model that is used to describe this process is that of the centre and periphery (Wallerstein, 1974). This presents the North as the centre, or 'core' of capitalism, and the South as its periphery. Through imperial conquest, it is argued, peripheral economies were integrated into capitalism, but on an inherently unequal basis. Supplying raw materials, which fed manufacturing industries in the core, peripheral regions became dependent upon foreign markets and failed to develop their own manufacturing bases. The infrastructure provided by colonial powers was wholly geared towards export; in many cases an economy might be dependent upon a single product. Dependency is thus:

> a continuing situation in which the economies of one group of countries are conditioned by the development and expansion of others. A relationship of interdependence between two or more economies or between such economies and the world trading system becomes a dependent relationship when some countries can expand through self-impulsion while others, being in a dependent position, can only expand as a reflection of the expansion of the dominant countries, which may have positive or negative effects on their immediate position. (Dos Santos, 1973)

Closely related to theories of dependency are those presenting the globe as a single interrelated system in which each country is understood in terms of its relationship to the whole. Immanuel Wallerstein's 'world system' (1974) and Worsley's notion of 'one world' (1984) are central to these ideas. It is from this context that notions of 'Third World' and 'First World' have developed; these terms explicitly recognise the way in which the world is divided into different and yet interdependent parts. The Third World, it suggests, is not natural, but created through economic and political processes.

Structures of dependency, the argument goes, are also repeated internally. Just as on an international level the centre exploits the periphery, within peripheral regions metropolitan areas attract the bulk of scarce local resources and services. They are occupied by the local elite, who, through their links with the centre, spend considerable time taking profit out of the country (by investing, for example, in costly education abroad). In a parallel process to international relations between centre and periphery, the elite also exploit surrounding rural areas through unequal exchange, for example in terms of trade between rural farmers and urban markets. Capital accumulation in the periphery is therefore unlikely to occur, both because of processes that suck capital into the metropolitan centre, and because of wider international processes which take it outside the country.

Dependency theory therefore understands underdevelopment as embedded within particular political structures. In this view the improvement policies advocated by modernisation theory can never work, for they do not tackle the root causes of the problem. Rather than development projects which ease the short-term miseries of underdevelopment, or support the status quo, dependency theory suggests that the only solution possible is radical, structural change. There are of course examples of this solution being followed. The radical internal restructuring of countries embracing socialism in the 1950s (China and Cuba are key examples) and the subsequent problems faced by them demonstrate that this is a route fraught with difficulty, however. Not only is state socialism often associated with extreme political repression, but, with the breakdown of communism in the Soviet Union and Eastern Europe in the 1990s, the new openness of China to world trade, aid and other manifestations of capitalism, and the economic crisis facing Cuba, its long-term viability appears limited.

The international political backlash against state socialism that gathered force during the 1980s has been matched by similarly forceful revocation of neo-Marxist analysis within academia. The generalisations of Marxist analysis, its inability to deal with empirical variation and its insistence on pushing all human experience into the narrow strictures of a single theory became seen as fundamental problems, and its explanatory framework too simplistic. The Marxian development critique was also attacked from within orthodox Marxism. Bill Warren argued that dependency theory failed to understand the nature of imperialism and capitalist development in the previously colonised South. Rather than remaining stagnant and perpetually underdeveloped, the ex-colonies were moving forward in a way largely in keeping with Marx's original ideas about the progressive (though destructive and contradictory) force of capitalism within his theory of historical materialism (Warren, 1980).

One of the main problems with dependency theory is that it tends to treat peripheral states and populations as passive, being blind to everything but their exploitation. While it is certainly important to analyse the structures which perpetuate underdevelopment, however, we must also recognise the ways in which individuals and societies strategise to maximise opportunities, how they resist structures which subordinate them and, in some cases, how they successfully embrace capitalist development.

Rather than offering solutions to societies in the capitalist world, dependency theory is in danger of creating despondency in its insistence that without radical structural change, underdevelopment is unavoidable. This does not mean that it has not had pervasive and continuing influence on developmental practice. It has contributed to the politicisation of development, which can no longer be presented as neutral. Internationally, this politicisation is expressed by the formation of alliances of Third World countries against the North, such as the Non-Aligned Movement, which since its inception following the Bandung Conference in 1955 has acted as a kind of international pressure group for Third World countries. Out of this emerged the Group of 77 countries (G77) that functioned as a counterbalance to the influence of the Northern industrial nations within the UN and its associated agencies (McGrew, 1992).

Notions of dependency have also contributed to, and reflect, the increasing politicisation of 'development' in the South at both grassroots and state levels. As an intellectual movement, its proponents were mostly situated in the South, in particular Latin America. Most fundamentally, neo-Marxist analysis raises a question largely ignored by theories of modernisation, but of crucial importance: who gets what from development? By focusing upon the ways in which profit for some is connected to loss for others, neo-Marxist analysis remains an important contribution to the understanding of development, even if as an analytical tool it is sometimes a little blunt.

While modernisation and dependency theory are politically polar opposites (one liberal and the other radical), they have a surprising amount in common. Both are essentially evolutionary, assuming that countries progress in a linear fashion and that it is capitalism that propels them from one stage to the next. Both emphasise the idea that change comes 'top-down' from the state, but they tend to ignore the ways in which people negotiate these changes and, indeed, initiate their own. Both are fundamentally deterministic and are based upon the same fundamental rationalist epistemology (Hobart, 1993: 5; Long and Long, 1992: 20). Most crucially for those at the receiving end of underdevelopment, neither offers a realistic solution. Modernisation's improvement policies, which wrongly assume 'trickle down' from profit-making elites to the rest, often do little to help the poorest and most vulnerable. Meanwhile the radical change suggested by dependency theory is often impossible to achieve.

In 2014 we can discern the influence of both modernisation and dependency theory in current practice and thinking. Notions of modernisation survive in much contemporary developmental thought. As we have already mentioned, agencies such as the World Bank remain committed first and foremost to promoting economic growth. Dependency theory also continues to influence thought and practice. It can be located, for example, alongside notions of empowerment that reject aid as a form of neo-imperialism, arguing instead that positive change can only come from within Southern societies. Paolo Freire's work on functional education, which has had a huge influence on some areas of developmental practice, in particular upon non-governmental organisations (NGOs), is an example of the practical application of neo-Marxist theory; first and foremost, he suggests, people need to develop political consciousness, and the route to this is through pedagogic techniques of empowerment (Freire, 1968).

Debates on gender and development have also increasingly involved awareness of the structural influences of global inequality and colonialism on gender relations, and of the need for women in the South to empower themselves rather than be recipients of Northern benevolence (Sen and Grown, 1987). More recently, the anti-global capitalism movement that we discuss at the end of the book is rooted, in some strands at least, in a neo-Marxist critique of the exploitation and inequality of capitalism and global finance. Other movements, though, such as the 'Bottom of the Pyramid' approach, or even Fair Trade, involve an inherent acceptance of neoliberal capitalism, heralding trade, consumption and the market (however reconfigured) as the solution to global poverty.

The demise of development theory

Despite these lingering influences, it was increasingly argued during the 1980s that the age of the 'grand narrative' was largely over. By the 1990s, neither modernisation nor dependency theory had survived intact as a viable paradigm for understanding change and transformation, or processes of poverty and inequality. There are various interconnected reasons for this. We have already suggested that neither theory can realistically explain the problems of global inequality and poverty. The strategies they offer for redressing such problems are also flawed. But there are wider factors operating too.

Politically, as the old polarities of the Cold War became obsolete after the 1980s, there was much talk of a 'New Global Order'. Although this concept was contested, the global and polarised struggle between the two opposing socioeconomic systems of capitalism and communism was clearly at an end. Francis Fukuyama's (1992) 'end of history' thesis was an assertion that Hegel's evolutionary view of history was no more and different forms of Western liberal democracy were now the only game in town. Increasingly it was no longer so easy to speak of the 'Third World', for the boundaries between the First and the Second, the communist bloc, were collapsing. Within the New Global Order at the beginning of the new century there was also no easy division between states on the periphery and those in the centre; the economic dynamism of Eastern Asia, for example, which is overtaking traditional centres of capitalism in North America and Europe, appears to disprove dependency theory. Combined with this, religious and ethnic revivalism, and the conflict with which both are often associated, have vividly indicated that understanding modernity is not nearly so simple a matter as was once assumed.

The 1990s: A time of 'postmodernity'?

By the 1990s, we were suggesting that anthropology and development were under the influence of 'the age of postmodernity'. While this term has various meanings, it is most simply explained as a cultural and intellectual rejection of modernity. Culturally, postmodern tendencies in the North could be traced back to the 1940s and 1950s, wherein the arts had increasingly moved beyond modernism to a broader, more pluralistic range of styles and techniques; by 1990 eclecticism, parody and multimedia forms were common. Likewise, the boundaries between 'high' and 'low' culture were increasingly broken down. Intellectually, postmodernism involved the end of the dominance of unitary theories of progress and belief in scientific rationality. Objective 'truth' was associated with the operation of power and was replaced by emphasis on signs, images and the plurality of viewpoints: there was no single, objective account of reality, for everyone experiences things differently. Postmodernism is thus characterised by a multiplicity of voices.

Postmodernism involved both conservative and subversive political tendencies. By insisting upon diversity and cultural relativity, it disregarded the possibility of common problems and thus common solutions. Revolutionary movements that advocated blanket remedies for social ills – such as state socialism – could not therefore be placed on the agenda. In its insistence upon locating particular voices and deconstructing what they say, however, postmodern approaches were inherently subversive. Edward Said's brilliant analysis of *Orientalism* (1978), for example, deconstructs Northern writings on the 'orient' to show how they homogenise and exoticise the 'East' and, by doing this, function as the ideological backbone of imperialism. Following Foucault, since the late 1970s and 1980s there was an increasing awareness of the relationship between discourse (fields of knowledge, statements and practice, such as development) and power.

From this, all categories that lump peoples or experiences together become politically suspect: by the time we wrote *Anthropology, Development and the Post-Modern Challenge*, the 'Third World', 'women' or the 'poor' were more often than not accompanied by inverted commas to show awareness of the problematic nature of such categories.

These arguments fundamentally undermined many of the earlier assumptions that came out of the colonial and postcolonial North about development. While not overstating the influence of an intellectual movement on those working within the aid industry, we suggested in 1996 that development theory had reached a profound impasse, and that this was partly a result of postmodern tendencies. Emphasis on diversity, the primacy of localised experience and the colonial roots of discourses of progress, or the problems of the Third World, radically undermined any attempt at generalisation.

To a degree, this was reflected in practice. As this short account implies, after the 1980s there were many different approaches to development, which – rather than being based upon one single theoretical creed, promising all-encompassing solutions in a single package – attempted to deal with specific problems. In the abandonment of generalised and deterministic theory, there was also an increasing tendency to focus upon specific groups and issues ('women', 'the landless'), a more reflexive attitude towards aid and development, and a new stress upon 'bottom-up', grassroots initiatives. These perspectives were already emerging in the 1970s, when stress upon 'basic needs', rather than macro-level policy aimed at industrialisation, was increasingly fashionable within aid circles. Instead of being radical these strategies were inherently populist. As part of a general trend that places people more directly on the developmental stage, they were closer to liberal ideologies of individualism, self-reliance and participation than Marxist ones of revolution or socialism. By the 1990s other trends included 'good government', human development and the increased use of cost-benefit analysis. Development, both as theory and as practice, had become increasingly polarised. While multilateral agencies such as the World Bank or United Nations agencies embraced neoliberal agendas of structural adjustment, free trade and 'human development', others increasingly stressed participation, empowerment, rights and the primacy of indigenous social movements. As the notion of development seemed to lose credibility, development practice became increasingly eclectic.

Just as postmodernist approaches problematised concepts and theories within development, they were also associated with a parallel crisis in anthropology (Grimshaw and Hart, 1993). While the degree of this is contested, there can be little doubt that, since the mid 1980s, many conventional tenets of the discipline were rigorously queried, both within and outside the professional establishment. To an extent, anthropology has always had some postmodern tendencies. Cultural relativism, one of the discipline's strong positions, insists upon recognising the inner logics of different societies. The world is thus presented as culturally diverse and composed of many different realities. What anthropologists had not tended to question until the 1980s, however, was the status of the knowledge that they gathered. Ahistorical generalisations based upon the observations of the 'objective' anthropologist were made in many of the classic ethnographies that served to disguise heterogeneities within the local societies under study. Theoretical frameworks such as functionalism and structuralism (which continued to influence some branches of anthropology up until the 1980s) tended to reduce societies to a series of commonalities, whether these be the notion of interdependent institutions which function to maintain the workings of the overall social system, as in functionalism, or the idea of common binary oppositions which underlie all social forms and to which all cultures can be reduced, as in structuralism.

In many ways then, anthropology's claim to represent and understand the diverse societies of the world was an easy target for postmodern critiques. One area in which it was attacked was the claim of so-called objective generalisation, or what Jonathan Spencer calls 'ethnographic naturalism' (1989: 153–54). This confers authority on the anthropologist by suppressing the historical specificity of the ethnographic experience. Given postmodern emphasis on local and diverse voices, the intellectual authority of the anthropologist who was supposedly providing an 'objective' account of exotic peoples was easily criticised.

Unease about the quasi-scientific paradigms of anthropology, and textual conventions which constructed anthropologist-authors as experts, was expressed by a series of publications over the 1980s, such as Clifford and Marcus's *Writing Culture* (1986), Marcus and Fischer's *Anthropology as Cultural Critique* (1986) and Clifford's *The Predicament of Culture* (1988). Writing conventions were not, however, the only problem. Growing reflexivity about the colonial heritage of anthropology and its contribution to imperialist discourses about the Southern 'other' contributed to increasing introspection concerning the subject's assumptions. Objectification of other peoples, we now realised, was linked to political hierarchy (Grimshaw and Hart, 1993: 8). Anthropological representations are not neutral, but embedded in power relations between North and South. This has led to what in feminist theory has been termed the 'politics of location' (Cornwall and Lindisfarne, 1994: 44–45) – the notion that one has no right to 'speak' for other groups, and the ascribing of legitimacy only to 'authentic' voices.

Since the late 1980s these arguments have led to various reactions. Some anthropologists moved away from ethnography and retreated into the analysis and deconstruction of text; others experimented with different styles of writing. A considerable number retained their interest in ethnography, but turned their attention to their own societies, or to others in the North. Rabinow (1986: 259) argued that one solution to the 'crisis of representation' facing anthropology is to 'study up' and research the powerful rather than the powerless. This might involve studying colonial authorities, planners, government – and development agencies. Connected to this was the call to 'anthropologise the West' (Rabinow, 1986: 241). Anthropologists, it was suggested, needed to turn their attention away from their exotic 'other' and focus instead upon their own societies. As we shall see in the rest of this book, the anthropology of development was to offer fertile ground for these new directions.

We shall return to what happened to the 'crisis of postmodernity' in anthropology at the end of this chapter. First, let us move development from the 1990s into the twenty-first century.

Global changes and continuities

As we noted in the Foreword, since 1996 many things have changed within the world of development. Global inequalities are worsening and wealth is increasingly concentrated, while the spread of capital and its interconnections are transforming much of the globe. Development reflects a changing balance of global power that includes the continuing growth of a set of middle-income countries that disrupt old distinctions between developed and developing countries, the post-9/11 securitisation of development, and a far stronger emphasis on the role of the private sector as a development actor. As a result, the 'field' for anthropologists of development now looks very different. In what follows we briefly outline these changes.

'Small d' development: changing global balance of power, unstable capitalism and rising inequalities

As outlined above, the idea of the 'Third World' has shifted over time. With its roots in the demarcation of colonial territories and later the developing world, the term was first associated with the French demographer Alfred Sauvy, who used it to refer to countries that were frozen out of the bipolar world of the Cold War era. He used it in a non-pejorative way similar to the way the third estate – the people – was used in the context of the French Revolution to draw attention to the need to bring those who had been marginalised to the centre. By the 1970s the term 'Third World' had come to mean something different – the idea of poor countries with third-rate governments, high levels of corruption and unstable environments. In the post-Cold War world, new euphemisms such as 'emerging economies' or 'the global South' continued to emerge, but with a continuing binary implied between rich and poor countries.

A simplistic notion of 'developed' and 'developing' countries has long been questionable. While accepting the reality of global inequality, long-standing relations of exploitation and the political value of 'Third Worldist' discourse, such a binary distinction obviously obscures important elements of diversity and complexity. By the 1980s, the rise of Japan as an economic powerhouse, the emergence of 'Asian tiger' economies, and the global power of oil-rich countries in the Middle East made the picture more complicated. As we note above, when the Cold War ended in 1989 the language that had set out a distinction between the 'First' World (the West), the 'Second' World (the Union of Soviet Socialist Republics [USSR], China and the Eastern bloc) and the 'Third' World (the rest) no longer fitted, even if large areas of Africa, Asia and Latin America remained impoverished. There were now many middle-income countries and emerging economies where the majority of the world's poorest people are now found and fewer poor countries. Shifts in global power render such over-simplifications even more unwieldy. The financial crash of 2008 and the continuing post-2008 economic uncertainty (including the euro crisis) that has infected previously powerful Western countries is beginning to change the way many see the balance of power in international affairs in important and far-reaching ways. In 2010 even the head of the World Bank Bob Zoellick was being quoted as saying that '2009 saw the end of what was known as the third world' (*The Economist*, 10 June 2010).

Systems of classification are changing in other ways. In the UK there used to be a clear distinction between NGOs that worked on domestic issues and those focused on poverty in 'developing countries'. This is increasingly challenged by organisations such as Oxfam and Save the Children Fund that have begun operating 'at home', establishing – sometimes controversially given their histories of assisting distant others – projects that aim to tackle poverty in the UK (Lewis, 2014a). Some developing countries are now also developers. For example, some developing country NGOs are internationalising. Bangladesh's BRAC has become the largest development organisation of its kind in the world, expanding from large-scale domestic work into development and humanitarian programmes in many other countries, becoming perhaps the first international NGO from the global South to do so. In Japan, Asia's first country granted full 'developed' country status, the post-2011 *tsunami* humanitarian and reconstruction effort raised important challenges for the third sector – and it was reported that the international sub-sector was able to carry out the work more effectively than the organisations of the domestic sub-sector.[2]

These kinds of changes help set the scene for a more geographically complex global landscape of countries, resources and relationships. The rise of China as a global economic power, for example, has dealt a fatal blow to the old Western binary worldview, and the gap between China and the group of Organisation for Economic Co-operation and Development (OECD) countries is rapidly closing. Today there is a veritable industry of economic commentators identifying trends, coining acronyms and predicting the changing global order. Most influential has perhaps been the BRICs (Brazil, Russia, India, China) idea, originated by Jim O'Neill at Goldman Sachs. Recognition that the developing world needs clearer disaggregating has also continued with various further spin-off terms such as the 'next eleven' (N-11) countries to follow the BRICs and, most recently, the MINTs – Mexico, Indonesia, Nigeria, Turkey.

The United Nations Development Programme's (UNDP's) *Human Development Report 2013 – The Rise of the South: Human Progress in a Diverse World* comments on the trend of growing interconnectedness and interdependence in the context of richer and poorer nations. The report suggests that 'the South needs the North, and increasingly the North needs the South ... The world is getting more connected, not less' (UNDP, 2013). While the economic progress made by China, Brazil and India is noted, the report identifies countries such as Mexico, South Africa, Indonesia and Turkey as growing in importance on the world's stage. The report also notes that each of these newly powerful countries appears to be following a unique and distinctive path that takes us a long away from the earlier certainties implied by the modernisation school of development that suggested that countries would, if they developed, follow a pattern drawn from a Western-derived blueprint.

The example of China, while showing how the global order is being transformed in favour of the 'world formerly known as Third' (as the Comaroffs [2012] have put it) is also illustrative of another important global trend – that of increasing intra-country inequality: 'within China, inequality is growing fast, and millions are relatively, if not absolutely, left behind in its headlong growth' (Sutcliffe, 2005). Rising global inequality is now a normal trend. In 2011 the OECD's report *Divided We Stand: Why Inequality Keeps Rising* drew attention to what it saw as an alarming rise in the disparities between rich and poor in the developed countries. This normally fairly cautious and even conservative 'club of rich countries' set out an important set of new data indicating that inequality was the highest for 30 years among its member countries. It explicitly challenged the logic of trickle-down theory that assumed benefits of economic growth would eventually reach those who were worse off. The roots of the growth of inequality were traced back to the deregulation of labour markets that promoted a rise in part-time and low-paid work, and the weakening of tax and benefit systems in many countries since the mid 1990s. Launching the report, OECD Secretary-General Angel Gurría called for governments to adopt strategies for 'inclusive growth' and warned that 'the social contract is starting to unravel in many countries'.

In January 2014, a report by Oxfam International on global inequality entitled *Working for the Few* (Fuentes-Nieva and Galasso, 2014) highlighted the ways in which rich elites have systematically captured economic opportunities through financial deregulation, tax evasion and other forms of self-interested political manipulation. The report presented an astonishing claim – that the world's richest 85 individuals own as much wealth (about £1 trillion) as the poorest half of the entire world's population of 7 billion. As Winnie Byanyima Oxfam International's Executive Director pointed out, this tiny elite group 'could all fit comfortably on a double-decker bus' (*The Guardian*, 20 January 2014). Around the same time, new attention was being focused on issues of inequality by the appearance of French economist Thomas Piketty's (2014) book *Capital in the Twenty-First Century*. Piketty argues that inequality is integral to capitalism and needs to be challenged by state intervention, which was a message that caught the mood of anxiety in Europe and the US around inequality and the book became an unexpected bestseller.

'Big D' Development: out with the old, in with the new?

The landscape of Aidland has always been one characterised by rapidly changing ideas, approaches and terms. A narrow emphasis on economic growth and technology transfer characterised the 1950s and 1960s, but by the 1990s the development world had begun to diversify and evolve, encompassing a wide range of alternative ideas, actors and models that now included NGOs, civil society, participation, empowerment and partnership. In the brief period of optimism after the end of the Cold War and before 9/11 it looked as if the alternative paradigms they embraced might gain transformative power and change the way development work was imagined and practised. As we shall see, this did not happen. Western aid and development became more heavily infused with the managerialist ideology that has gained ground across many of the traditional donor countries and the global context in which international aid has traditionally operated is in the process of undergoing dramatic change.

Development was originally presented in terms of the creation of economic growth and the transfer of new technology that would bring material benefits through modernisation. Such thinking remains in the mainstream, continuing today in the priorities of international agencies. For example, since the 1980s the IMF and the World Bank have placed a strong emphasis on 'structural adjustment' reforms that have aimed to liberalise markets and reduce the role of government. The World Trade Organization (WTO) locates development within the reform of international trade regulations and the freer movement of capital between North and South.

While official development institutions such as the World Bank have become principal advocates of what Harvey (2007) has called the 'neoliberal way', there is also a range of alternative or counter-discourses within the worlds of international development that seek to question, modify or challenge mainstream ideology and institutions. For example, the dominant emphasis on economic growth and gross domestic product (GDP) as an indicator of progress is increasingly under challenge from those who emphasise a set of broader ideas about quality of life, wellbeing and happiness. The government of Bhutan has famously rejected GDP as an indicator of progress and put in its place a system of gross national happiness (GNH) that places an emphasis on social spiritual, physical and environmental wellbeing.[3] Economics in the twentieth century moved away from an earlier emphasis in the eighteenth and nineteenth centuries on happiness – acknowledged for example by Adam Smith as the 'ultimate human goal' – and became exclusively focused on material goods (Thin, 2009). Both in relation to development and in society more widely, a renewal of interest in wellbeing has taken place, including among anthropologists (such as Thin, 2009; Matthews and Izquierdo, 2009). As we will discuss later, alternative ideas about development have also emerged in the form of promoting participation, empowerment and citizenship rights.

One major issue that makes Aidland profoundly different from twenty years ago is the strengthening of managerialist practice among aid agencies. In contrast to the values of alternative development discussed earlier performance measurement and audit have now become key preoccupations. As Mike Power (1997: 142) has shown in his analysis of the 'audit society', there has been a steady momentum in UK and other Western societies towards the reshaping of public life around 'evaluation, assessment, checking, and account giving' that raises some critical questions about 'programmatic ideals of "performance" and "quality" and the technologies through which they are made operational'. For example, it is common for new arrangements to be put in place based on the need for taxpayers to know that their money is spent economically, efficiently and effectively (the three 'Es') and to challenge 'cosy cultures of professional self-regulation'. The problem with this, as Power (1997) argues, is that while few people would argue with the logic of seeking to improve public accountability systems, the result is usually the introduction of technical solutions that are better at signalling that efforts are in place than actually contributing to the desired change.

Such pressures have been increasingly felt not only within the domestic contexts of Western societies but also in their development agencies. Within the new 'arts of government' that are integral to most forms of neoliberalism (see Ferguson, 2009) aid has come to be seen as far more about managing things than about understanding people. The world of development aid has begun to face 'increasing internal and external critique, and a growing need to show results for tax payers' money' (de Haan, 2009: 173). This brought for example a new emphasis on performance indicators, most noticeably in the form of internationally agreed targets for poverty reduction. The UN Millennium Development Goals (MDGs), agreed by 189 nations in September 2000, put in place

8 poverty reduction goals and 18 targets, including halving poverty and hunger, creating universal primary education, and halting and reversing HIV/AIDS. The stated objective was to reduce by half the number of people living on less than US $1 per day by 2015.

The MDGs were followed by the aid effectiveness agenda set in motion at the OECD Paris Declaration meeting in 2005, which set out a set of new principles for the relationship between donors and recipient governments based on *harmonisation*, *ownership*, *alignment*, *mutual accountability* and *results*. The 2008 Accra Agenda for Action built on these principles further, adding themes that included commitments to improved planning and predictability within aid processes, more use of country systems to deliver aid rather than donor systems, less rigid conditionality, and the relaxation of tied aid restrictions (Mawdsley, 2012). The 2011 OECD's Fourth High Level Forum on Aid Effectiveness (HLF-4) met in South Korea to review progress and agree priorities for the future, based on the two continuing priorities of implementing the Paris aid reform principles, and the new mantra of improving 'value for money' through investing aid resources more cost-effectively. The outcome was the Busan Partnership for Effective Development Cooperation that, for the first time, agreed a framework for development cooperation that encompasses traditional donors, those involved in South–South cooperation, private funders and the BRICs, and civil society organisations.

Their new preoccupation with value for money in relation to aid provision reflects a developmentalist disposition that now emphasises more than ever before the roles of markets and business as essential to development. For example, the NGO sector that was once seen as animated primarily by an alternative 'non-profit' ethic that set it apart from the for-profit world of business now increasingly embraces the latter's principles, tools and ideologies. The British NGO Save the Children was once known for campaigning against international corporations involved with child labour and infant formula, but it now embraces a partnership with the pharmaceutical company GlaxoSmithKline as a way of funding health workers and new vaccines (Lewis, 2014b). Alongside the trend for closer working relationships between NGOs and the private sector is the rise of corporate social responsibility (CSR) as we discuss in Chapter 3 and the Conclusion.

With market principles increasingly introduced into development work, changes have continued to take place in the way that aid is delivered and assessed. For example, 'payment by results' is a new form of contracting in which the commissioner of a service only pays the provider once 'a pre-determined result has been achieved and independently verified' (Eyben, 2013: 15). Another trend is the increasing use of experimental research techniques to compare intervention approaches and measure impact as part of the managerialist rhetoric of 'evidence-based policy-making'. For example, Banerjee and Duflo (2011) in their bestselling book *Poor Economics* have popularised the use of randomised control trials (RCTs), in which the results of a project or policy intervention are compared against a control population or location.

Providing resources, managing them effectively and securing measurable 'results' are the new priorities in the delivery of development assistance. In their book *A Perfect Storm*, Tina Wallace and her colleagues (2013: 4) describe:

> a culture of managerialism where change must first be envisaged, then detailed, described and planned for. Once implemented, projects must demonstrate the achievement of pre-set results, which must be measured and reported on in quantitative terms. Change is understood as linear, logical and controlled, following theories of change based on a cause-and-effect model.

For anthropologists, all of this has raised important questions: who was setting these targets and why? How could quantitative targets capture important issues of access, service quality, exclusion and power? What are the risks that perverse incentives emerge that would draw attention away from poverty and local struggles in favour of easy wins? Furthermore, the application of such experimental research methods as RCTs (developed originally in the field of drugs trials and medical research) to development policy settings has led some anthropologists to argue that this contributes to a narrowing of definitions of what counts in terms of knowledge and evidence, and represents the antithesis of earlier alternative 'participatory' ideologies and approaches (Lewis, 2014).

The earlier promise of building a set of alternative or non-mainstream development approaches is increasingly lamented as having failed to reach its potential. The political space for thinking about development has become ever more constrained, as Wallace et al. (2013: 16–17) highlight in their study of what they see as development policy's current conditions of a 'perfect storm' that combines managerialist donor thinking on logic models, results, targeting and value for money, an increasing role for the private sector and its values, and the increased professionalisation of the development NGOs sector. As they point out:

> INGOs started out, in the main, as independent players in development and humanitarian aid, working with local partners and communities ... concepts such as solidarity and accompaniment were common in many organizations, and the space for critiquing and challenging dominant norms existed, though it was often not a comfortable space to occupy, something experienced by many gender staff, for example ...

Another set of important changes since the mid 1990s are the geopolitical shifts in the foundations of Aidland. Western aid is in decline and new or re-emerging country donors are playing larger roles in a 'multi-polar' international development. Emma Mawdsley's (2012) book on the new aid donors shows that power is beginning to drain away from the established apparatus of post-1945 development through which the European and North American nations transferred foreign aid, technologies and policies to poor countries via the Bretton Woods institutions of the World Bank and IMF, the UN system and the bilateral and non-governmental development agencies. The global financial crisis that began with the 2008 banking crash has reduced the capacity (or willingness) of many Western governments to provide international aid.[4]

Most prominent among newcomers is Chinese aid, which in 2007 provided an estimated US $18 billion to countries in Africa, and has prioritised simple infrastructure building and agricultural training over governance or social development. Countries such as the Gulf States, South Korea, South Africa, Russia, Poland, Thailand, Turkey and the Czech Republic have each established their own aid programmes, which are viewed as important domestically and badge of status and soft power globally. There has also been an expansion of new private sector development philanthropy, in the form of agencies such as the Gates Foundation with an endowment of more than US $30 billion. Finally, remittances as international resource transfers from rich to poor countries began to outstrip foreign aid in many contexts. For example, in Bangladesh, a country previously viewed as overwhelmingly dependent on foreign aid was receiving three times its aid as remittances from migrant workers by the end of the 1990s (Lewis, 2011a).

The more powerful countries among the new aid donors such as China and Brazil have begun to speak the language of South–South cooperation, partnership and self-reliance, and contrast an anti-imperialist stance with the traditional, conditional aid approaches and charity of the former colonial powers. Aidland is now in flux, as the new aid entrants offer welcome alternatives for recipient countries. Some remain suspicious of the new donor motives and fear that they conceal similar forms of self-interest to the old ones, and that the gains achieved by international aid in some settings might now be put at risk (Mawdsley, 2012).

Finally, in the period that followed 9/11 a trend towards the 'securitisation of aid' has returned Aidland to the paranoid era of the Cold War, in which development activities were seen as a bulwark against a hostile 'other' (see Chapter 2). International development assistance has once again come to be seen unapologetically by governments as an instrument of foreign policy and 'soft power'. Vast amounts of aid have flowed to Afghanistan, Iraq and Pakistan as part of the Western response to the 2001 attacks on the United States that President George W. Bush termed 'the war on terror'. Overlaid upon the neoliberal framework of development assistance that had been growing in strength during the 1990s, the securitisation agenda has had widespread effects and changed the priorities of international development assistance. It has for example ushered in a new 'othering' and problemati-sation of Muslim societies, organisations and remittances; diminished the independence of humanitarian aid agencies; and constricted the available 'civil society space' in which development alternatives can be imagined and discussed (Howell, 2006: 126).

New forms of capitalist penetration, transformation and resistance in developing countries

Neoliberal policies have led to the restructuring of welfare systems through the reduction of the role of the state, the increased marketisation of service delivery and the loosening of arrangements to regulate international flows of capital in both North and South since the 1980s. Central to this was the imposition of structural adjustment policies (SAPs) upon many developing countries in the 1980s. At the start of the twenty-first century, levels of global inequality remain profound, with 15% of the world's population controlling most of the world's assets and setting the key rules around economic activity. Formerly of the International Labour Organization (ILO), Guy Standing (2009) writes of the emergence of the 'precariat', a new global class across both North and South left vulnerable by neoliberal policies, whose existence is characterised by short-term jobs, an absence of reliable welfare arrangements, and an unstable and increasingly unregulated economy. Broadening the focus of the anthropology of development requires us to focus on the international political and economic relationships that underpin wider global political economy, which is contributing to a changing balance of power between different parts of the world, to increasingly uneven concentrations of poverty and deprivation, and to rising inequalities.

Today the Eurozone crisis is also bringing the experience of painful economic adjustment and associated political instability to many people who live in countries previously seen as wealthy and stable. The extent and extremes of poverty in areas of Asia, Africa and Latin America cannot be compared to disadvantage and inequality in the industrialised societies of 'the North'. But poverty is increasingly being experienced by many individuals and social groups within societies previously regarded as wealthy, as global social inequalities have increased within countries as well as between them. While the scale of poverty is obviously different, basic causes and processes are similar, and the interconnections are more and more visible. For example, John Gaventa's (1999) study of community-level organisations in the Appalachian communities of the United States and among workers in Mexico, and the structural interdependence of exploitation and exclusion within both communities, led him to argue for the existence of 'Norths in the South' and 'Souths in the North'. This inter-connectedness is increasing as migration, displacement, trade, conflict, and transnational institutions grow and intensify (Kothari, 2005).

Anthropologists have long been concerned with understanding the position of marginalised people and with implicitly or explicitly supporting their struggles. In *Confronting Capital*, which deals with anthropological engagement with the spread of capitalist development across the world, Gardiner Barber et al. (2012: 2) discuss the ways that 'episodes of collective dissent' increasingly characterise a global system riven by deep tensions and oppositions signified, for example, by the Arab spring, the Occupy movement and Wisconsin's citizen support of the assault on public sector unions. They make the case that 'anthropologists have unique critical purchase on the complexities and contradictions of these and varied other responses to capitalism, its ever changing faces' (2012: 2). The political economy tradition within anthropology can be revitalised to usefully interrogate the contemporary world, and in particular to engage with the 'entanglements of ordinary people within capitalism' (2013: 2). They frame such an analysis within what they term a 'double agency', in which both the anthropologists doing ethnography and the people that they write about are engaged in confronting capital in different but complementary ways. Such work builds on the Marxist tradition in anthropology (such as the work of Maurice Godelier and Sidney Mintz), feminist scholarship (Kathleen Gough and Louise Lamphere), and Marxist historians such as Raymond Williams, E.P. Thompson and Eric Hobsbawm. It encompasses both the long-standing tradition of anthropological work on agrarian livelihoods and contestations over power and resources to also include a focus on how people more widely are making daily decisions about how to engage with markets and capital.

In a related vein, a recent collection of writings edited by Mathews and Vega (2012) focuses on what they call 'globalization from below'. This is seen as a process of transnational flow of people and goods that involves relatively small amounts of capital and primarily informal transactions, and which takes place in both the 'developing' and the rich countries of the North. The authors contrast 'high end globalization' involving corporations with 'low end' traders moving used or copied merchandise across borders, the use of containers and luggage to sell goods in the informal sector, and street vendors working through personal connections and with wads of cash, and without lawyers or copyrights. An anthropological approach such as this offers two levels of analysis: it provides large-scale ethnographic descriptions of the 'routes, nodes and channels' that structure this globalisation from below, and provides highly intimate portrayals of the different actors involved such as small entrepreneurs, traders and peddlers.

In this sense, globalisation from below is an idea and an approach that can be seen as a contestation of the types of development policies usually propagated by the World Bank, IMF and WTO in the form of 'a direct confrontation with the macroeconomics and development policy advocated by international financial institutions' (Mathews and Vega, 2012: 11). The tendency is for these organisations to promote one-size-fits-all rules for developing economies, along with neoliberal policies that favour corporations, and this is seen by Mathews and Vega as ultimately contributing to greater global economic inequalities. Yet this form of globalisation from below might be seen to follow certain free market principles as well since it seeks to evade state control. In this way the rules of neoliberalism remain in place, but rather than merely serving the status quo act in opposition to it: 'as the effort to sustain local practices and local communities as against the ravages of "rich countries' capitalism"' (2012: 11). Perhaps, with echoes of Ferguson's (2009) arguments, Mathews and Vega argue that this tendency can therefore be seen as 'a warmer, more human form of neoliberalism' that 'does not necessarily sever human social bonds' in the ways other more rigid framings of globalisation sometimes suggest. Such a perspective also rejects the simple formulaic development policy prescriptions that have become popular among those who argue that such grassroots informality and mobilisation will eventually vanish as development problems are 'solved'. Such arguments are to be found in Hernando de Soto's *Mystery of Capital* (2000), which prescribes changing the laws and reducing the barriers to property ownership, and in the ideas of the Grameen Bank's leader, Muhammad Yunus, that institutions for microcredit will help to reduce poverty.

Anthropology and post-development: into the twenty-first century

When we wrote *Anthropology, Development and the Post-Modern Challenge* in 1996 it was partly in response to Arturo Escobar's attack on anthropologists working in development, whom he accused of failing to react to changes taking place within anthropology, questionable methodological practices and – most damningly – for reproducing discourses of modernisation and development (1991: 677). In a later work he suggested that development made anthropological encounters with Third World others possible – just as colonialism once did. Rather than challenging it, anthropologists 'overlook the ways in which development operates as an arena of cultural contestation and identity construction' (1995: 15).

As we shall show in what follows, anthropologists engaging in development, both academically and more practically, have in large part met these challenges head on. Indeed, more generally postmodernist critiques that we outlined earlier have in many ways strengthened the discipline. Rather than retreating into lit-crit or becoming overly solipsistic, since the late 1990s anthropologists have continued to do what they do best: studying the everyday worlds and cultures of ordinary people across the globe, revealing realities that are otherwise largely ignored, and by so doing gaining alternative ways of seeing that destabilise conventional orthodoxies. As those anthropologists researching 'globalisation from below' that we discuss above show, as well as the recent growth of interest in the anthropology of 'policy worlds' (see Shore et al., 2011), such work has resonance outside anthropology as well as within it.

While anthropologists have largely withdrawn from creating grand narrative theories of human society or culture – perhaps because they have lost confidence in the efficacy of sweeping generalisations – the basic task remains the same: cross-cultural comparison, rooted in qualitative, long-term fieldwork. New methods such as multi-sited ethnography have been largely accepted, as has the endeavour of 'studying up'. Likewise, reflexivity – the placing of the anthropologist into his or her text and reflecting upon their authorial and subjective role in creating their knowledge – has become commonplace.

Contemporary global issues are in many instances now framed as problems of D/development – and are increasingly mainstream or even central concerns of the discipline. As we shall also see, the study of development institutions and policy has become a fertile area for anthropologists wishing to study up. It is also a way in which the discipline can move beyond the silencing of identity politics to a more politically engaged praxis. Just as some feminists have argued that there must be postmodern 'stopping points' rather than endless cultural relativism (Nicholson, 1990: 8), and that one such point is gender, as we suggested in 1996, another is the politics of poverty.

What, then, do we mean by development? We use the term here to refer to processes of social and economic change that have been precipitated by economic growth, and/or specific policies and plans, whether at the level of the state, donor agencies or indigenous social movements. These can have either positive or negative effects on the people who experience them. Development is a series of events and actions, as well as a particular discourse and ideological construct. We assume that these events and actions are inherently problematic; indeed, some aspects of development are actively destructive and disempowering.

Rather than promoting development per se, what we aim to do in this book – as with the 1996 edition – is not only to trace the history of anthropology's relationship with D/development, but also to show how the social and political relations of poverty and inequality might be better understood, and thus combatted, through the generation and application of anthropological insights. We define poverty as a state in which people are denied access to the material, social and emotional necessities of life. While there are 'basic needs' (water, sufficient calorific intake for survival and shelter), many of these necessities are culturally determined. Poverty is, first and foremost, a social relationship, the result of inequality, marginalisation and disempowerment. It occurs in the North as well as the South (although much of our attention in this book will be confined to the South). We suggest that while we need to move beyond the language and assumptions of development, the application of anthropology in attempting to construct a better world is as vital as ever in the twenty-first century.

2

Applying Anthropology

Since the earliest days of the discipline, some anthropologists have been interested in using – or 'applying' – their knowledge for practical purposes. Applied anthropology has been used as a term covering a wide range of roles and activities. Raymond Firth (1981) identified four main roles. First, applied anthropologists can undertake client-oriented research either as commissioned academics or as professional consultants. Such clients have included professionals from business, public administration, health care services, education provision, social work, defence, agriculture and, of course, international development. Second, they can perform a mediation role between members of a community they are studying and outsiders and try to interpret local culture and issues to outsiders. Third, anthropologists can try to contribute to the formation of public opinion on issues relating to a small-scale community, through journalism or participation in other media, and thereby try to influence public policy. Finally, they may participate directly by helping to provide assistance to people they are studying during times of crisis.

More recently, applied roles have been seen as further diversified. In *Applied Anthropology* (1993), John van Willigen has listed a total of 14: policy researcher, evaluator, impact assessor, needs assessor, planner, research analyst, advocate, trainer, culture broker, expert witness, public participation specialist, administrator/manager, change agent and therapist. Applied anthropology, sometimes also termed 'practising' anthropology, therefore came to be seen as the 'field of enquiry concerned with the relationships between anthropological knowledge and the uses of that knowledge in the world beyond anthropology' (E. Chambers, 1987: 309).

Alongside other social science disciplines, anthropology has long been viewed as a tool that might help policy-makers, administrators and managers understand (and therefore influence or control) people that they were dealing with, whether employees in an organisation, consumers in the market place, or members of a local community. Some applied anthropologists remained inside academia and worked from time to time on applied research projects or as consultants. Others chose or were compelled by the lack of university job opportunities to operate outside academic settings altogether. For example, 'corporate anthropology' has become a growing field in the US, where private companies today employ 'thousands of anthropologists' to do market research or human resource management (Welker et al., 2011: 57). Rejecting a strict separation between knowledge and action, applied anthropologists have engaged on the basis of both formal contractual and voluntarist principles with worlds beyond academia.

Application has been controversial from its earliest days because it raises difficult philosophical, political and ethical questions. When research is linked to action, or undertaken on the basis of a commercial contract, how does this affect a researcher's objectivity and the quality of their work? When anthropological knowledge is put at the service of commercial or military interests, can the anthropologist remain true to the discipline's traditional stance of privileging the perspectives of the powerless and the marginalised? Applied anthropology has come to be viewed as suspect by many within the discipline because of 'its ties with colonialism, its questionable linkages with Cold War machinations, and its supposed complicity with those who create, rather than solve, social problems' (Rylko-Bauer et al., 2006: 179). Other critics have simply argued that applied anthropology, if it is judged in terms of results, has rarely been effective in influencing wider policies and practices (Bennett, 1996). For all these kinds of reasons, many in the academic establishment have regarded applied anthropology as a rather dubious offshoot of the discipline.

This chapter reviews the history of applied anthropology, outlines the different roles in which applied anthropologists have worked, both in their own and in other societies, and discusses the ethical and other issues that arise from these. We then consider the particular ways applied anthropology has played a part within the specific field of international development. Finally we consider and question the long-standing tendency to distinguish between applied and academic (or 'pure') anthropology – and argue instead that priority should be given to building a critically engaged, ethically grounded form of public anthropology that can transcend this unhelpfully dichotomous way of thinking.

Anthropology, cultural relativism and social change

Nineteenth-century anthropologists were engaged in debating two major sets of theoretical issues that bore directly on the practical application of anthropological knowledge. The first was around the notion of change itself. Within anthropology, social change was initially debated between diffusionists (such as the German *Kulturkreise* ['culture circle'] school, which included Fritz Graebner and Martin Gusinde), who saw change as gradually spreading across cultures from a common point, and evolutionists (including Lewis H. Morgan and Herbert Spencer), whose ideas rested on the assumption that all societies, if left alone, would evolve through broadly similar stages. In time the diffusionist arguments, which recognised that cultures interact with each other and are thereby altered, gradually replaced those of the evolutionists.

With the growth of functionalism in the twentieth century, anthropology began to concern itself more with the means through which societies maintained themselves than with the ways in which they changed.[1] During the 1930s, the functionalist perspective of modern British social anthropology, represented by the work of Arthur R. Radcliffe-Brown and Bronislaw Malinowski, emphasised the relationships between different elements of a society in the ways in which it reproduced and maintained itself. The functionalists eschewed history, and paid very little attention to how communities changed over time. The tendency to study societies as if they were static remained strong in the period up to the Second World War, but was challenged by anthropologists interested in what was termed 'culture contact' in the colonial territories.

Yet one of the key figures advocating an applied approach was Malinowski himself, who had first made the case in 1929 in an article entitled 'Practical anthropology' for anthropology to engage with the world beyond academia (Malinowski 1929). He gradually moved away in his work from the tribal economics of exchange systems and suggested (in an appendix to his 1935 book *Coral Gardens and Their Magic*) that he had betrayed his ethnographic principles by omitting details of links his islanders had formed with colonial officials, missionaries and traders (Hann and Hart, 2011). He had deliberately focused on peoples' 'traditional' activities. Identifying with the idea of indirect rule used within the British Empire, he then argued in *The Dynamics of Culture Change* that it was now the duty of anthropologists to make themselves 'useful' and 'to contribute as much information and advice as possible to the controlling agents of change' (1961 [1945]: 12). Earlier classic ethnographic monographs such as Malinowski's *Argonauts of the Western Pacific* (1922) and Evans-Pritchard's *The Nuer* (1940) had been ahistorical functionalist accounts that had largely ignored issues of change. But gradually anthropological work was beginning to take fuller account of the shifting histories and context of communities and seeking explanations for social and political change. Increasingly, change came to be seen as inseparable from society itself. As Beattie (1964: 241) pointed out:

> Change is taking place in all human societies all the time. Sometimes it is sudden and catastrophic, as when a system of government is destroyed by revolution and replaced by a different one; sometimes it is gradual and hardly perceptible, so that even the members of the society themselves scarcely notice it.

Anthropologists therefore adapted their discipline in the 'late colonial era to studying transformations resulting from colonial encounters' (usually referred to by Malinowski as 'culture contact') (Hann and Hart, 2011: 45). This prepared the way for anthropologists to become involved in studying what was later called 'development'.

The second major theoretical issue debated by these early anthropologists was cultural relativism, which formed an obstacle that stood in the way of ideas about the application of anthropology, particularly in the United States. The rise of the 'cultural anthropologists' (whose outlook drew on the relativist ideas of their founder Franz Boas) gradually displaced evolutionary ideas about culture after the First World War. In contrast to the evolutionists, who saw social change in terms of culture's adaptation to the environment, Boas's work among the Inuit/Eskimo had led him to adopt a view of culture as being completely independent of 'natural' circumstances. As Bloch (1983: 126–28) has argued, the view of culture held by these anthropologists led to the predominance of a cultural relativism that held that 'it is wrong to evaluate one culture in terms of the values or knowledge of another'. The dominance of cultural anthropology in the 1950s squared with prevailing American political ideas. While recognising the existence of cultural differences, cultural relativism made possible the coexistence of different ethnic groups within one society, at the same time justifying non-interference by the state in people's lives.

Relativism therefore raised important problems around the ethics of intervention by anthropologists in the communities in which they worked, a dilemma that can never be satisfactorily resolved. If a culture was to be understood on its own terms, as Ruth Benedict's *Patterns of Culture* (1934) had convincingly argued, what business did members of one culture have telling those of another what to do? The tendency to move anthropology towards direct engagement was a challenge to the values and ethos of relativism, as Eric Wolf (1964: 24) suggested:

> Applied anthropology, by definition, represents a reaction against cultural relativism, since it does not regard the culture that is applying anthropology as the equal of the culture to which anthropology is to be applied.

The implications of these debates are still being felt among many anthropologists in academic departments around the world: between those who favour a more open-ended theoretical development of the discipline, and those who, crudely speaking, might see anthropology as a tool for social engineering or, as we ourselves might prefer to put it, are trying to contribute their knowledge to address issues of poverty, inequality and rights for marginalised sections of the world's population.

The evolution of applied anthropology

Although its emergence was mainly contemporaneous, applied anthropology evolved differently in Britain and the United States. In Britain, the roots of applied anthropology lay – like the discipline itself – in colonial power, while in the United States its origins were bound up with state policies to control Native American populations. In the US, a Bureau of American Ethnology had been set up in 1879 as an arm of federal government with the aim of providing research for policy-makers. In Britain, the first use of the term is associated with Lane Fox Pitt-Rivers in 1881 (Howard, 1993: 369).

The British colonial system imposed structures and institutions that profoundly influenced the societies, politics and cultures of Asia, Africa and beyond. Not surprisingly, both colonial administrators and some anthropologists were interested in the possibility that research might play a role in assisting the colonial administrations. Arthur R. Radcliffe-Brown became one of the best-known early advocates of this idea. After his appointment as Professor of Social Anthropology at the University of Cape

Town in the early 1920s, he established courses in 'applied anthropology'. One of Radcliffe-Brown's main motivations was the reduction of conflict between the white and black communities in South Africa, and he emphasised a potential role for anthropology in contributing to better cultural understanding between communities (Kuper, 1983). From this period onwards, some anthropologists gained fieldwork opportunities and funding within the British colonial system, usually in African territories, where they worked on issues such as reforms to local land tenure systems, managing succession within tribal authority structures, labour migration and customary law.

Some anthropologists were commissioned to undertake specific research on prescribed areas of government interest, while others provided information on an informal basis. Anthropologists also contributed to the training of colonial civil servants. Much of the work was part of the effort to manage the British policy of 'indirect rule' in many parts of Africa and Asia, in which day-to-day affairs were left in the hands of 'traditional' local institutions and rulers by the colonial authorities and overseen from a distance. Anthropological engagement was driven by different motivations, ranging from critical support for colonial administrations to insider attempts to subvert the system from within. Not surprisingly, applied anthropology became controversial both among colonial administrators and anthropologists. Anthropologists, particularly those with liberal or anti-imperialist views, tended to view local, non-Western culture as something to be defended, almost at all costs, against colonial power. From their side, colonial administrators tended to see anthropologists as impractical, other-worldly academics with little of value to contribute to the day-to-day administrative problems of the territories. It is still a matter of dispute as to how far anthropologists influenced colonial policies: much evidence suggests that probably they did not (Kuper, 1983).

By contrast, applied anthropology in the United States mainly took shape in domestic rather than international arenas. Although Margaret Mead and Robert Redfield worked overseas during the 1920s and 1930s, most anthropologists were at home, busy documenting the ruined cultures of Native Americans, whose communities provided them with opportunities for fieldwork 'in their own backyards' (Wolf, 1964: 13). There was in fact a 'triple origin' to US applied anthropology: work with the administrators of Native American reservations (the closest parallel to British experiences with colonial administration), engagement with research on the social and cultural aspects of industrial organisation at the Harvard Business School, and official research studies of US rural communities sponsored by Roosevelt's Bureau of Agricultural Economics and Rural Welfare (Bennett, 1996).

Opportunities for anthropologists to get involved in applied roles increased when the 1934 Indian Reorganization Act was passed by the US Congress. This had the aim of giving the Office of Indian Affairs better access to local information as it tried to increase Native American participation in the management of their own economic affairs and reverse resource depletion on their lands. An Applied Anthropology Unit was set up in order to look into the creation of self-governing bodies, manage settlement patterns on newly acquired lands, improve education policies, build local morale and make better use of existing local institutions to bring about 'economic rehabilitation and social control'. In a political strategy that had parallels with the British policy of indirect rule, the aim was for this applied research to inform policy and practice under the new Act, based on 'the necessity of taking persistent Indian attitudes into account in planning for their social and economic adjustment to dominant American values' (H.G. Barnett, 1956: 37). In the late 1930s, the Bureau of Indian Affairs also embarked upon a large-scale natural resource survey with the Department of Agriculture in which anthropologists also played a role.

Relationships that were created between applied anthropologists and policy-makers in the context of Native American affairs also began to influence the mainstream of US anthropology. For example, the term 'acculturation' was coined by US anthropologists to explain how 'groups of individuals having different cultures come into intensive first-hand contact, with subsequent major changes in the original culture patterns of one or both groups' (Haviland, 1975: 366). This idea led anthropologists to examine change in terms of contacts between cultures, which led to such new ideas as 'syncretism', where old features blended with the new, or 'deculturation', where aspects of culture were lost altogether. Acculturation was a useful concept in that it provided anthropologists with a framework for analysing change, but it also contained crucial limitations. Cultural change was presented mainly in terms of the reorganisation of different components across cultures. Emergent aspects of culture, as well as the more subtle changes in relationships between different institutions, tended to be given less consideration. The emphasis on first-hand contact also overlooked the tremendous power of the media to influence culture without the need for any direct contact.

Applied anthropology also flourished in the business sector during the first part of the twentieth century in the US. In one famous example, managers at the Western Electric Company commissioned a series of studies at its Hawthorne Works in Chicago. These examined the relationship between working conditions and productivity, and brought in Harvard psychologist Elton Mayo to help analyse what appeared to be complex and contradictory results. Mayo was aware of anthropological methods through friendships with both Malinowski and Radcliffe-Brown. He brought in one of the latter's students W. Lloyd Warner for a new stage of experiments that began in 1931, known as the Bank Wiring Observation Room studies. This phase moved the research from formal interviews to an anthropological 'participant observation' approach, in order to focus on what workers actually did rather than simply on what they said. This generated influential new insights into the importance of informal social relationships and interaction among the workforce, including the idea of the 'Hawthorne effect' in which workers were found to perform better when researchers took an interest in their working conditions.

The New Deal era in the United States that followed the Great Depression led some anthropologists to work with agricultural scientists in the Soil Conservation Service during the 1930s and the 1940s, though such work was generally small scale. Charles Loomis, for example, worked in Peru on issues related to the establishment of an agricultural extension system (Kedia and van Willigen, 2005). The American Society for Applied Anthropology was founded in 1941 (far earlier than any comparable body in Britain) and began publishing a wide range of articles in its quarterly journal, *Human Organisation*.[2] Loomis was among its founders.

There was an intensification of anthropological engagement with various types of applied work during the Second World War. When the United States entered the war in 1941, the government made extensive use of professional anthropologists and as many as 90 per cent may have taken part in war activities (Mead, 1977; Price, 2011). Some anthropologists worked in areas occupied by US forces, such as the Trust Territory of the Pacific Islands, and were charged with facilitating the cooperation of the local population with the authorities in organised activities such as construction work. Many of the founders of British anthropology, such as Evans-Pritchard and Edmund Leach, had also worked with the government during the war.[3] In an echo of British colonial approaches, training was given to military officers and administrators in anticipation of future roles administering territories taken from the enemy (H.G. Barnett, 1956: 12).

In one particularly shameful episode, some anthropologists were involved in the US government's mass internment of Japanese Americans after the attack on Pearl Harbour in 1941. Fearing the threat of espionage and collaboration with the enemy, the government moved more than 100,000 people from the West Coast into camps located in the US interior, though none was ever charged with any crime. Some anthropologists, such as Marvin Opler, worked with the authorities but took a critical stance, co-authoring a hard-hitting book entitled *Impounded People* in 1946 (Opler et al., 1969).

Post-war applied anthropology

After the Second World War, Britain's slow process of decolonisation was beginning to get under way and anthropology began to withdraw from its remaining colonial links, losing a major source of applied funding. The British Foreign Office was reorganised during the 1950s and anthropology's official influence in the new emerging postcolonial world faded, although informal links doubtless continued between anthropologists who had been involved with government earlier. When the new Overseas Development Ministry was established in 1964, there were no anthropologists involved (Grillo, 1985: 16). At home, there were some anthropologists who briefly developed new applied links with the world of business managers, but these soon petered out (Mills, 2006).

In both the UK and the US, a divergence between mainstream academic anthropology and applied anthropology had gradually formed. This had promoted a feeling among many university-based staff that only the second-rate anthropologists carried out applied work, while the 'real' anthropologists worked on loftier, self-determined and theoretically informed subject matter. The status of applied anthropology within the discipline as a whole remained low, and it was widely seen by those in university departments – according to Lucy Mair (1969: 8) – as an 'occupation for the half-baked'. The discipline was also out of favour with the new nationalist leaders of the former European colonies, who tended to identify anthropology with the old order.[4] Nevertheless, the emerging field of international development did become a specialisation within applied anthropology, known as 'development anthropology' (Epstein and Ahmed, 1984).

Despite a loyal commitment to applied anthropology among small numbers of anthropologists, the field remained small until a resurgence took place in the late 1960s and early 1970s. This was a period during which the university sector was undergoing expansion and more academic jobs were being created. Some socially concerned anthropologists were beginning to reject the confines of a purely academic job and sought to apply anthropological knowledge to the important domestic social issues of the day. For instance, during this period there were anthropologists who became involved with the new policy concerns around what were termed 'race relations' (Grillo, 1985: 2). Anthropologists provided a set of ideas about how recognisable physical differences between peoples can be manipulated symbolically by those wishing to exploit social, economic and cultural differences (Beattie, 1964: 271).

In the UK, several initiatives were under way to promote wider public engagement by anthropologists from the 1970s. New applied professional associations were established, including Social Anthropology and Social Work, and the British Medical Anthropology Society. The Group for Anthropology and Practice (GAPP) was established in 1981. During the 1980s, an ongoing attack on higher education by the Thatcher government led some within the discipline to defend anthropology by asserting its academic purity. Yet teaching jobs and research openings for trained anthropologists within the university system were few. The search for opportunities outside academia for working anthropologists became a priority for those recently qualified anthropologists looking for work and. once again, tensions between pure and applied anthropology surfaced.[5] GAPP became a refuge for those wishing to engage outside academia, and it evolved into the British Association for Social Anthropology in Policy and Practice (BASAPP) in 1988, finally becoming known as Anthropology in Action (AinA) in 1993. Its journal *Anthropology in Action* and an associated website continue to function. In 2004, a new network of practising anthropologists was formed within the Association of Social Anthropologists (ASA) and this has gone some way to overcoming long-standing distrust between the two groups (Wright, 2006).

In the United States, many anthropologists and other social scientists perceived the Second World War as 'a lesson in cultural dominance on a scale never seen before' (Wolf, 1964: 14). The dubious activities of some anthropologists during the war now undermined the legitimacy of applied work among many academic anthropologists, tarnishing its reputation. Social scientists, including anthropologists, generally became warier of government and its interventionist foreign policy.

Anthropologists responded by withdrawing from involvement in wider social issues through their work, retreating towards a more strictly delineated arena of 'academic' ethnographic and theoretical research. For example, while Mead and Redfield had pursued practical anthropological concerns with domestic food and nutrition studies during the 1940s, there was a general move away from applied concerns in what Montgomery and Bennett (1979) describe as a 'return voyage to tribal ethnology and theoretical interests' that generally carried anthropologists away from applied concerns (cited in Rhoades, 1984: 3). At home, the field of 'business' anthropology had also dwindled away by the next generation, due to the decline of the human relations management school, the arrival of more university job openings for anthropologists during the baby boomer era, and a set of new American Anthropological Association (AAA) ethics guidelines that restricted work to only that which could be freely disseminated, making it very difficult to do applied anthropology in commercial settings (Baba, 2005). As in Britain, university departments expanded and many practising anthropologists took the opportunity to enter academia and gain 'respectability'.

Applied anthropology became a relatively small specialist interest, often relegated to second-tier status by anthropologists in the elite universities. Some new doors opened during the 1970s for the revitalisation of applied anthropology, particularly in the development field. Organisations such as USAID began recruiting anthropologists, particularly those who had developed new methodological interests around the use of surveys, quantitative work, and impact assessment and evaluation. Anthropologists contributed to debates about the limitations of 'trickle-down' theory and lobbied for more resources to be channelled to low-income groups. Nevertheless, the role and impact of anthropologists on development theory and practice was marginal (Hoben, 1982: 356). Other forms of knowledge about development, particularly the new science of development economics, held more sway than anthropology.

Domestically, there were occasional calls for more public engagement by anthropologists with the wider world of policy. Cyril Belshaw's book *The Sorcerer's Apprentice* (1976) advocated closer ties with policy-makers, arguing that local anthropological knowledge was an important and under-utilised resource that could inform and sensitise planners. Belshaw elaborated an anthropological concept of 'social performance' that he argued could be used as a tool to evaluate the effectiveness of a social system in delivering goods, services and 'satisfactions' in the eyes of its people. But anthropology itself remained largely rooted within the academic establishment. In the US, it was primarily based within liberal arts colleges as opposed to science campuses, isolated from the practical concerns of economics, management and agriculture.

Chicago anthropologist Sol Tax was an exception. Tax developed the concept of 'action anthropology' from his involvement with the Fox Native American community, emphasising an activist approach that aimed to move beyond the confines of both academic and applied anthropology, and pursue a responsibility to the members of a community side by side with the acquisition of knowledge (Polgar, 1979: 409). According to Tax, an anthropologist undertaking action anthropology has two goals: 'to help a group of people to solve a problem, and ... to learn something in the process' (Blanchard, 1979: 438). As well as allowing for the explicit involvement of the anthropologist in community problem solving, this approach emphasised the need for the anthropologist to present his or her findings to both the academic and the 'native' community.

As national security issues became prioritised during the Cold War, the dilemmas of engaging with power periodically fuelled ethical debates within the discipline. Attempts were frequently made to recruit anthropologists into the service of foreign policy, and they were particularly seen as having the potential to contribute to counter-insurgency work (Hoben, 1982). In 1964, an initiative called Project Camelot was announced and involved the largest social science research grant ever offered by government. Initiated by the US army's Special Operations Research Office (SORO), it was designed to develop a model for understanding and managing internal conflict in developing countries, starting with the case of Chile (Belshaw, 1976). There were clear links with dubious US foreign policy objectives: Project Camelot caused controversy in academic circles and was widely discredited. Only one anthropologist was involved and the project itself 'died a quick death' (van Willigen, 1993). This interest in anthropology and counter-insurgency was to re-emerge strongly again in the twenty-first century in the context of the so-called 'war on terror', as we will see later on.

In other parts of the world, different traditions of applied anthropology emerged. In India, the traditional concerns of anthropology with minority or 'tribal' communities (as they remain known) led to the institutionalisation of anthropology after 1947 within the newly independent state. Anthropological texts formed part of the training given to Indian civil servants, just as it had in colonial times. Anthropology was seen as having a specialised contribution to make in the task of national social and economic development, and a government Department of Anthropology established in 1948 became a Central Advisory Board for Anthropology in 1958, charged with furthering the economic development of the 'tribal' areas. Nevertheless a distrust of anthropologists' motives continued in some quarters of Indian society, where they were (not without evidence) suspected of being more interested in keeping 'tribal' people 'in a zoo' than in helping to address their real problems (Mathur, 1989: 43). In the continent of Africa, another 15 years or so of colonial government had to be endured before anthropology began to find a place within newly independent countries.[6] In Mexico, anthropology was also 'institution-alised as public policy' through various national-level institutions that were designed in the late 1940s to promote the use of anthropology as a tool for national integration (Sittón, 2011).

In some countries, anthropologists were not attracted to the idea of application. In France, Baré (1997) explains that there has generally been far less interest in applied anthropology than in 'pragmatist' countries like Britain or the United States:

> For many French anthropologists, the very idea of 'applying' a 'science' in a definite way is an absurdity, since research cannot know what it can find; it cannot be 'piloted' through questions that are not of its own. If people want to use anthropology, they should just read anthropology books.

The sub-field of applied anthropology has therefore had a patchy history, and its impact overall has been relatively small. Anthropology came to see itself – in Raymond Firth's memorable phrase – as the 'uncomfortable discipline', in the sense that it continued to shun the mainstream and remained more concerned with the challenging of established ideas and values, and exposing underlying interests and contradictions. Two areas of discomfort were experienced: the first by managers and administrators who usually found that anthropologists questioned their ideas and practices, and the second among anthropologists themselves, as they struggled to maintain an appropriate distance from the policy regimes that affected the communities they were trying to understand (Wright, 1995). Anthropologists were often awkward types, prone to take an oppositional position in relation to authority, or an isolationist stance with little effort to participate in public engagement or forge links with other professionals.

Yet, as we saw earlier in the discussion of 'Orientalism', anthropological work nevertheless fed into the way colonial power represented and made societies governable.

As we saw in Chapter 1, anthropology began to look inwards from the middle of the 1980s. It embarked upon a period of critical re-evaluation and reflected on issues of representation and textuality, based on a critique by Clifford and Marcus (1986). A new postmodern anthropology began concerning itself with the need for a reflexive approach to ethnographic writing, relegating the concept of practice to the back burner. However, postmodernist debates among sociologists and political scientists about the politics of the rise of 'new social movements' also began to feed into anthropology and fuelled reflections around power, politics and representation. The realisation that much of applied anthropology had been taking place within what Escobar (1995) calls the 'dominant discourse' began to stimulate discussion about anthropology's potential to challenge its hegemony and to draw attention to other, less visible discourses.

The rise of postmodernism impacted upon applied anthropology by problematising its boundaries with the rest of the discipline, and assumptions about theory and action. Johannsen (1992: 79) argued that it suggested new, more radical approaches to applied work, such as the reinvigoration of Sol Tax's tradition of action anthropology, and that such an approach could provide:

> an infrastructure for sustained self-reflection by the people being studied, which will ultimately produce a process of self-assessment. It aims at empowering people by providing a context that better enables them to represent themselves, their culture and concerns.

Pels and Salaminck (1991, cited in Mills, 2006) correctly problematise the idea of a simple clear boundary between 'pure' research and what they call 'other ethnographic practices' including applied anthropology. Where such a boundary has been asserted, this may be understood as being primarily in support of academic anthropology's periodic interest in maintaining a 'comfortable boundary of professionalism' around itself.

Development as an applied field of anthropology

So far in this chapter we have explored the general landscape of applied anthropology, and touched upon the domestic and international contexts in which it emerged. As we saw in Chapter 1, the increased importance of international aid and the growth of development agencies after the Second World War made it an obvious area to attract the interest of applied anthropologists. Development, as we have seen, was an idea that related both to interventions in the 'Third World' and a broader universal concept relating to economic growth, sustainability and justice. As with other arenas of applied anthropology, some anthropologists practised development anthropology as university-based academics doing contract work, while others joined development agencies to work in specific roles as evaluators or advisers. Anthropological knowledge, with its emphasis on detailed, fine-grained, qualitative, people-centred data came to be viewed by development organisations as a potentially valuable commodity.

By the 1990s many anthropological professionals had become involved as 'insiders' in mainstream development in broadly supportive roles, while others took more independent or oppositional 'outsider' roles in which they have attempted to mediate between development agencies and local communities in ways they hope will radically alter or challenge the terms of engagement. Other anthropologists engaged with development indirectly: by conducting research that seeks to understand how people experience development, either in terms of what is done to them as part of planned interventions and policies, or how they themselves go about attempting to shape the process of building a better life. As we saw in Chapter 1, development was also open to different meanings and interpretations – for example, as both 'immanent process and an intentional practice' (Cowan and Shenton, 1995: 28), or an issue involving communities in rich countries as well as poor ones (Salisbury, 1977) – making development a field that was attractive to anthropological interest in both applied and non-applied ways.

Anthropological engagement with development is not therefore easily positioned within these applied/non-applied categories. Indeed, we argue that while 'development anthropologists' have gone some way towards forming a distinct identity for themselves during recent decades, this is a distinction that needs questioning. The applied/pure dichotomy cannot be intellectually sustained within anthropology more widely since all anthropologists are practising and applying knowledge at some level, are part of society and, through their representations, are changing it. Equally, within the field of development we cannot neatly break down categories of activity into practice, policy and theory. Instead, as we shall argue below, anthropological work on development is better approached holistically and critically.

In the section that follows, we explore briefly some of the applied roles anthropologists have taken in the field of development work with agencies (that is, 'big D' development), before moving on to those undertaken more broadly as engagement with 'little d' forms of unfolding change.

Projects

After the Second World War, projects became central to the organisation of mainstream development activity, as they had to other areas of the world of planning. Projects were time-bound structures for an intervention based around a series of staged activities, known as the 'project cycle', involving a series of stages from identification to implementation, completion and evaluation. As Robertson (1984: 121) puts it, projects are where everyone involved – from outside expert to local farmer – have what is expected of them on a day-to-day basis 'known, scheduled and quantified' by a 'single administrative authority'. Development, whether conceived as large-scale infrastructural work such as the building of a dam or bridge, or as 'softer' areas such as health or education provision, became organised into a series of projects often designed in a top-down way by international development donors with little consultation with local communities.

Anthropologists were employed by development agencies to help with project design, appraisal and evaluation. Some were invited to carry out 'impact studies' among the local community to assess whether or not the project's objectives have been met. Sometimes these studies were combined with academic, longer-term research concerns in familiar cultural contexts, while others were 'one-offs', where anthropologists found themselves in unfamiliar settings. Anthropological work was able to open up important new thinking on projects. For example, Richard Salisbury's (1977) study of the social impact of a hydro-electric project in Quebec commissioned by the government and the James Bay Development Corporation was influential. The study had been commissioned only after anthropologists had complained that insufficient attention had been given to the proposed project's impact on the local environment and population. Salisbury identified what he termed a 'prism of perceptions' among the different interest groups involved:

> Every different interest group concerned with the Project, government, industrial, or private, Indian, or white ... had a stereotypic perception

of what 'development' was, and then interpreted what it learned about this project in the light of that perception ... (1977: 171)

These perceptions included development 'as heroism' (in the context of Quebec's assertion of itself as a modern state managing capital and energy), 'as national self-realization' (as assertion of francophone Canada's nationalism), 'as jobs' (in the context of the employment of construction workers and expert consultants), 'as civilisation' (a view emerging from the missionary history of Quebec, in which the forest frontier is moved back and its inhabitants brought into the mainstream of society and, finally, 'development as disaster' (held by environmentalists, anti-capitalists and those who romanticised what they regarded as the 'traditional' way of life of the local Cree Native American communities) (Salisbury, 1977).

The idea that economic and social change could be framed within projects was problematic for anthropologists, but central to the top-down, controlling urge of development. Projects were artificial constructs in which the idea of 'social engineering' was dominant. They were framed as rational plans, but quickly became complicated during implementation by the complex behaviours of real people around which straightforward decision-making boundaries could not easily be drawn (Pottier, 1993). For example, this was apparent in the case of the large integrated rural development projects (IRDPs) introduced by the World Bank and the United Nations during the 1970s. IRDPs aimed to take a comprehensive approach to tackling problems on several sector fronts simultaneously – for example, agriculture, health care provision and education – so that each component could be linked in one large project. Despite attempting to supplement conventional planning methods with a measure of community participation, most remained heavily top-down and organised through large bureaucracies that proved themselves incapable (or unwilling) to involve local people in decision-making (Black, 1991).

Lucy Mair (1984) recounted a series of hair-raising stories of planners foisting inappropriate projects on hapless rural people, which include resettlement schemes where people are moved without adequate compensation and the introduction of new technology resulting in economic benefits being captured by men within the household at the expense of women. Mair's was essentially an optimistic view of the potential of anthropology to render development more people-centred, and she reassures us that: 'if I concentrate on the disasters, it is because they are what anthropological knowledge might help to prevent on later occasions'

(1984: 111). Another anthropologist who argued that anthropology could make projects more people-centred was Ward H. Goodenough (1963: 17), who saw the way forward for community development in the Third World as moving from 'attempts to reform others' to 'helping others reform themselves'. In his book *Cooperation in Change*, he explored the idea that the 'cultural ignorance' of planners and project staff was often to blame for the failure of development projects, and that the role of anthropologists was to help put this right. He viewed anthropological insights as providing a set of new principles for building 'cross-cultural cooperation'.

Such work created a continuing set of ethical dilemmas for anthropologists. They were sometimes viewed (or presented themselves) as the representatives of local people and simply asked to provide certificates of social acceptability for projects, in place of undertaking direct consultation with people themselves. Sometimes anthropologists were seen merely as 'fixers' who could be brought in when things went wrong, rather than being involved in consultation, design and implementation from the start. Little wonder that Robertson described the role some development anthropologists found themselves in as 'pathologists picking over project corpses', with little involvement in planning (1984: 294).

By the 1980s and 1990s, project failure and critique enabled new alternative 'people-centred' development paradigms to emerge, at first within the more activist-oriented NGOs whose profiles had grown during this period. Some NGOs in both North and South had become sites for new critical development thinking. These ideas included empowerment, participation, gender equality, indigenous knowledge and sustainability, all of which moved development thinking away from being about mainly infrastructure, technology and economic growth closer to a community-level approach with which many anthropologists felt comfortable. The work of radical participation specialist, Robert Chambers and many others around 'rapid rural appraisal' and 'participatory learning and action' became highly influential, and was directly linked to anthropological insights. In some ways there were parallels between the experiences and ideas of NGO field staff going into local communities to try to understand local needs, and the ethnographic tradition of anthropologists undertaking participant observation; the two began to feed off each other in the bourgeoning interest in participatory work (Schuller and Lewis, 2014; Long and Villareal, 1994).

Within the mainstream official agencies, these trends influenced the rise of 'social development' as a specialised form of expertise that was informed by anthropological thinking. By the late 1980s the Department for International Development began expanding its recruitment of 'social development advisers', many of whom were anthropologists by training (Rew, 1985). A social development adviser's job was to ensure that social issues were accounted for during the design and implementation of development interventions. In some cases this literally meant asking 'How does this affect the local population?' In others, community development strategies were added or the new ideas of gender and empowerment developed into training courses for officials. The World Bank's *World Development Report 2001* (World Bank, 2001) was also now promoting approaches that placed greater emphasis on ideas of 'empowerment', echoing a growing demand within development agencies for 'social analysis' to be provided by non-economist social scientists. Another set of pressures came from governments increasingly intent on ensuring that publicly funded research could be shown to be 'useful' and policy 'relevant' (Paiement, 2007).

By the late 1990s the situation had changed again. As we saw in Chapter 1, as international aid has become more managerialist and technocratic, the management of aid moved 'upstream' and the centrality of projects began to wane. Projects became characterised as unsustainable, isolated and uncoordinated, and new aid management tools emerged to provide aid more directly to host governments in order to fit into national policy frameworks. For example, donors worked with governments in more coordinated ways through consortia to develop poverty reduction strategies into which, it was decided, aid could be funnelled in a more coordinated and effective way. The mechanism of 'budget support' also fitted into this new paradigm, with donor funds channelled directly into national budgeting processes. The idea was to avoid different donors prioritising different, sometimes contradictory objectives by forming a common 'pot'.

As we also saw, the Millennium Development Goals adopted in 2000 represented a new framing of the objectives of development into measureable time-bound targets to reduce poverty. This was to be measured at the macro level, so there was less interest in micro-level processes at the level of projects and local people. Attention to household-level livelihoods and gender inequalities began to fade, and official donors and development agencies recruited fewer specialists in these subjects. These changes meant that fewer anthropologists were being employed and, in particular, agency demand for 'social development' expertise began to recede. Many of these trends were further consolidated by the

2005 Organisation for Economic Co-operation and Development Paris Consortium meeting on aid effectiveness. A new set of 'aid effectiveness principles' was agreed, based around ideas of improving 'coordination' of development work between donors, the 'harmonisation' of formerly disparate development objectives, and efforts to promote recipient country government 'ownership' of mutually agreed poverty reduction goals and approaches. A focus on impact was now leading development agencies back to forms of economic cost-benefit analysis expertise and experimental research such as randomised control trials that could generate 'evidence' of intervention efficacy.

Advocacy

Advocacy has long been central to development anthropology, where anthropologists felt a duty to mobilise or speak out on behalf of the disadvantaged groups they studied when the interests of local communities were threatened. The role of the anthropologist as advocate and mediator informed Lucy Mair's book *Anthropology and Development* (1984). Mair made the case for the anthropologist's role as intermediary between 'the developers' and 'the developed', acting as go-betweens between the top-down developers and voiceless communities. If a development intervention is to achieve its objectives, she suggested, anthropologists have a responsibility to become involved to try to ensure that certain kinds of problems are avoided.

Anthropologists have helped establish or been drawn to organisations such as Survival International, the International Work Group for Indigenous Affairs (IWGIA), and Cultural Survival, concerned with the efforts of 'indigenous' people to gain more control over their lives (Escobar, 1992). These efforts have contributed to forms of 'committed anthropology' that extended beyond the formal academic career to try to bring to public attention cases of genocide and ethnocide, campaigning on human rights abuses and lobbying for material help for communities under threat (Polgar, 1979: 416). At the international level, resettlement issues also became a major focus for advocacy work by applied anthropologists. Thayer Scudder became a key player in persuading authorities to pay attention to the needs and rights of relocates, in what became known as 'resettlement anthropology' (De Wet, 1991). Such forms of anthropological advocacy have tended to be more associated with resistance to outside interventions rather than with prima facie agenda-building: for example, supporting opposition from local communities to the building of a dam, or the preservation of local culture in the face of change or repression.

At the same time, the tendency for anthropologists to speak on behalf of other groups was problematic. Some of the pitfalls of advocacy were exemplified by the work of Oscar Lewis, who, in research in a slum in the 1950s in Mexico saw himself as both a 'student and a spokesman' for the poor, who (it was assumed) were unable to speak for themselves. The publication in Spanish of Lewis's book concerning the 'culture of poverty' in a slum in Mexico (*The Children of Sanchez*, 1961) caused a political storm and he was accused by the government of having insulted the culture of the people of Mexico (Belshaw, 1976). More honest perhaps was Richard Salisbury's (1977) response when working on the James Bay hydro-electric project in Canada:

> When I was asked by whites what the Cree Indians wanted (expecting a simple answer, 'a maintenance of the traditional way of life') I could only reply: 'I don't know. They have conflicting wants.'

These days, anthropologists are now more likely to see themselves as facilitating and supporting action by disadvantaged groups, rather than simply speaking on behalf of them. At the same time, some anthropologists have also worked to problematise relativist positions around the tensions between universal rights and culture, and trying to ensure that 'more attention is paid to empirical, contextual analyses of specific rights struggles' (Cowan et al., 2001: 21).

During the 1990s, NGOs became a central focus for anthropological work of this kind, providing vehicles both for outsider advocacy (defence of those marginalised by development) as well as insider advocacy (community-based development work). There had been relatively little interest in NGOs in either the academic or practical worlds of development before this time. But by the end of the decade – in keeping with the rising tide of non-state privatising imperatives associated with the new neoliberal development policies – the idea of the NGO had almost come to define what development was all about (Lewis, 2005). NGOs received a higher share of international development funding, often on contract from governments, and their overall profile increased in the Western media. More importantly, NGOs became an important source of employment, both for middle-class managers and field workers in the developing world, and for researchers – including anthropologists – looking for jobs and research contexts. Anthropologists working within NGOs began to contribute both to community-level advocacy and to higher-level policy advocacy work.

Dilemmas of applied work

Anthropologists engaging in applied work have been subjected to considerable criticism by their purer peers. This has centred on four basic points: quality, objectivity, ethical integrity and opportunism.[7]

First, the practical pressures of applied work are seen as leading to compromises in its quality. The tradition of ethnographic work is long-term immersion, often taking several years. Commissioned research will often require the anthropologist to work quickly within a narrow time frame and produce findings – and sometimes direct recommendations – that may oversimplify complex issues and questions. While this may be possible in contexts with which the anthropologist is already familiar through earlier work, it may also lead to compromised or 'quick and dirty' work in which too many corners have been cut. For John W. Bennett, applied anthropology is 'not a genuine academic field' (1996: S49), but should be seen instead as 'a set of opportunities that some people with anthropology degrees pursue out of hunger or genuine social dedication or both' (1996: S48).

Second, there may be a loss of objectivity if research is carried out by an academic 'for hire' in a government-sponsored or commercial setting. Some take the view that researchers engaged to undertake contract work become mercenaries who are unable to 'speak truth to power'. Such work may be 'tainted' by employment interests and may simply provide credibility to, or assist directly with dubious policies. Others argue against either/or views, that it is better 'to have a seat at the table' and that, for example in the case of anthropologists working with corporations, that:

> by [us] being there, applied anthropological work in the corporate context may be presenting us with the opportunity to foreground a less oppositional but neither fully complicit stance ... (Cefkin, 2011: S26)

Such debates go back to the colonial era. The utilisation by anthropologists of opportunities for fieldwork within colonial administrations had long been subject to criticism. Talal Asad (1973) mounted a powerful retrospective attack on their aims and motivations, and indeed upon anthropology itself, which he rightly argued was created through an unequal encounter between Europe and the rest of the world. This asymmetric history provided the West with the opportunity to gain access to the types of cultural information upon which anthropology depends. Anthropology was complicit in this act of domination, even though anthropological work did not necessarily reflect the ideology of colonial administration.

Today there are increasing incentives and pressures on anthropologists to work with businesses, either directly within specially created posts in marketing or corporate social responsibility, or through research links between companies and universities. As Coumans (2011: S40) suggests:

> Academics who engage the private sector are increasingly rewarded by their academic institutions and by academic grant makers who are interested in private-sector partnerships.

Companies in the mining sector, for example, increasingly seek specialists – including anthropologists – to address issues of social and environmental harm or oversee 'corporate social responsibility' and 'ethical business'. Whether or not this is merely a form of public relations for such companies is, of course, a matter of debate in which anthropologists are widely engaged.

At the same time, attempts by powerful governments – and the US in particular – to use anthropological knowledge for instrumental purposes continue to fuel debate about the engagement of anthropology with wider public affairs. Since the colonial period, questions about the politics of academic engagement have regularly resurfaced – from the Second World War and the Cold War to the US intervention in Vietnam and more recently in Iraq and Afghanistan. In his book *Weaponizing Anthropology* (2011), David Price documents the ways that war and conflict have always helped to shape the work that anthropologists do. He discusses the way that, alongside conventional military activity, the US military has regularly tried to engage in 'cultural warfare', drawing on what they see as the discipline's capacity – through the method of ethnographic work – to get close to other societies, so that this information can be used for informing counter-insurgency work. During the post-9/11 era, as international aid has once again become more closely linked with security interests, the debate has once again become heated. For Soederberg (2004) development has now come to be seen as a foreign policy tool to 'pre-empt' instability and terrorism.

This raises crucial questions about the ethics of involvement. In the late 1960s, there were several scandals around anthropological involvement in US counter-insurgency activities in South East Asia and in Latin America which led to fierce debate and conflict at the AAA and the production of guidelines for ethical professional conduct. The 1969/1971 Code of Ethics was based on the idea that anthropologists should 'do no harm' to their research subjects, drawing from the Hippocratic oath that is used by medical practitioners (Price, 2011). The inadequacy of professional standards became apparent during the 2000s, when the AAA once again became concerned with anthropology's collaboration with the military. For example, in 2007 news of the US military's new Human Terrain System (HTS) project caused considerable concern among anthropologists. In Afghanistan the HTS project, according to Alexander Star in *The New York Times* (18 November 2011), aimed to:

> engage in 'rapid ethnographic assessment' – conducting interviews and administering surveys, learning about land disputes, social networks and how to 'operationalize' the Pashtun tribal code.

The idea was to draw upon anthropological knowledge and 'embed' anthropologists in military units in Iraq and Afghanistan. The AAA issued a statement that described the ways that the HTS contravened its Code of Ethics. As a result, most academic anthropologists chose not to become involved. According to Star, paradoxically, many subsequent anthropological writings find fault with the international effort in Afghanistan 'for its lack of ethnographic insight'.

In the development policy field, ethical dilemmas were complex. In 2004, David Mosse experienced a set of ethical tensions between research and development expertise in relation to his ethnography of a rural development project in India, on which he had worked as a researcher and consultant over a 10-year period. This work was later published in the acclaimed book *Cultivating Development* (2005) (discussed in more detail in Chapter 3), but for a while it looked as if the study might never be released because of objections raised by some international technical consultants and project managers (though not, interestingly, in the main by field staff or funders). After being invited to read and comment on the draft manuscript, a small number of these individuals complained formally to the university, the publisher and academic association in an attempt to prevent the book coming out. Writing later about the experience, Mosse (2011b: 50) reflects on the surprise he experienced when a non-evaluative anthropological account that mainly presented a picture of a 'worthwhile' project that benefited thousands of India's poorest people could encounter such resistance. The reason, he suggests, is that those opposing his account saw it – from their perspective, perhaps correctly – as potentially damaging to their professional reputations and therefore livelihoods. It reinforces and reveals a potential arrogance at the heart of the ethnographer who attempts to represent lives of others and 'denies others their cosmopolitan claims' by rendering them 'local' (2011b: 63). In this type of ethnography, 'the users of research are also the subjects of research', who may challenge the assumed primacy of academic knowledge, since it competes with other expert knowledge systems.

There is still a high level of ambivalence at the heart of anthropology about ethics (Campbell, 2010). While well intentioned, the available ethical guidelines continue to be seen by many anthropologists as inadequate and often impractical. The AAA regularly returns to the question of revising its ethical guidelines without ever satisfactorily resolving the issues. The UK Association of Social Anthropologists has similarly grappled with the issue.[8]

In such debates, applied anthropologists can arguably be seen to have led the way, producing the first guidelines around ethics in the early 1940s (van Willigen, 1993). David Price (2011) suggests that the concept of 'dual use' might be useful for social scientists. This idea is drawn from biological sciences, where research on something such as viruses for benign purposes may be used by those with malign intent, making it necessary for researchers to think through from the start the wider implications of their research.

There is also the criticism that policy-makers who commission applied work rarely take much notice of the results anyway – particularly when a study's findings are unwelcome, complex or critical. Public officials have long found anthropologists awkward to deal with. Adam Kuper (1983: 107) quotes colonial administrator Sir Philip Mitchell (Governor of Uganda 1935–40, Fiji 1942–45, and Kenya 1944–52), who wrote that anthropologists behaved as if:

> they only were gifted with understanding, busied themselves with enthusiasm about all the minutiae of obscure tribal and personal practices [from which studies] resulted a number of painstaking and

often accurate records ... of such length that no-one had time to read them and often, in any case, irrelevant, by the time they became available, to the day-to-day business of government.

Compared with engineers and economists, anthropologists lacked status. Evans-Pritchard, in an article written in 1946, bemoaned the fact that in 15 years of work in the Sudan he had never once been asked his opinion by the authorities. H.G. Barnett (1956: 49) has suggested that: 'No matter how tactfully it is phrased, the truth is that anthropologists and administrators do not, on the whole, get along well together.' These difficulties surfaced particularly in the case of anthropologists working in association with government agencies, where prejudices, preconceptions and doubts on both sides tended to make attempted collaboration a rather marginal endeavour.

Anthropologists were also seen as awkward, having gained a reputation for being over-concerned with the intellectual independence of their academic agendas, and as being unrealistically inhibited about the dangers of 'selling out'. This tendency was visible, for example, in the case of agriculture, where anthropological engagement has been surprisingly low compared with, say, the discipline of agricultural economics, which has benefited from the 'client relationship with society' pursued by the economics establishment (Thurow, 1977, cited in Rhoades, 1984: 4). Furthermore, anthropologists who did attempt to influence policy on the basis of their research tended to find that they had very little influence in practice and quickly became disillusioned. The resultant pessimism about such matters owed more according to Erve Chambers to:

a failure to distinguish between policy science and the making of policy – the former being an area in which most anthropologists have, largely by choice, had little experience and impact. To become involved in policy decision-making is not the same thing as becoming responsible policy scientists. It implies learning a good many other skills than those associated with scientific enquiry. (1987: 320)

Bemoaning the failure of planners and practitioners to take notice of research and advice offered became a common trope within anthropological writings about development. For example, Richard Salisbury (1977: 171) describes his belief that if an anthropologist simply communicates facts to the planners this would be fed into the design and implementation process of the project: 'we soon found that facts were ignored unless they confirmed prior perceptions, while non-facts were accepted if they were in confirmation of perceptions'.

Finally, there is the criticism among some academic anthropologists that applied anthropology is a profession that operates in a 'parasitic' way in relation to the 'pure' academics upon whom they depend for ideas that they then apply outside academia. The allegation here is that applied anthropologists consume and profit from mainstream work, but fail to contribute anything back to the discipline (Wright, 2006: 33). Such allegations are contested by anthropologists such as Paul Sillitoe (2007), who points to the value of the wider discipline of applied research fields, such as the field of indigenous knowledge that has contributed to theoretical developments in the discipline, and to theory about the application of knowledge. And, as we have seen, applied colleagues have historically pushed for clarifications around ethical codes of practice and this too has informed the mainstream discipline on an issue that is important to all concerned.

Engaged anthropology in the twenty-first century

Let us now move beyond the traditional, somewhat narrow definitions of the field of applied anthropology and consider the renewed enthusiasm for a public anthropology that engages at the level of practice, and at the level of ideas. The 'anthropology of development' is a broader category than the applied term 'development anthropology'. It includes work on the ethnography of development organisations and institutions, as well as critical work on development as a system of ideas. Most importantly, the new anthropology of development needs to move behind narrow definitions of 'development' and its specialised institutions to encompass a new vision of development as the wider struggle by people and governments to combat poverty and inequality. In this section, we examine three strands of engaged anthropology that increasingly take us beyond the confines of both 'applied' and 'development' concerns: 'public anthropology', 'activist research' and 'protest anthropology'.

Public anthropology

Today, old debates about 'applied' versus 'pure' anthropology have become increasingly outmoded. Instead, there is renewed interest in wider forms of anthropological engagement. This shift is driven by multiple factors, such as the tradition of the 'public intellectual' with an obligation to contribute to wider society, as well as offering a response to the increased pressures on academics to engage with the world around them. In the UK, the government now requires universities to produce information not only on research quality but also on its 'impact'. Recent years have seen promising new efforts within anthropology to try to connect with audiences beyond academia. For example, *American Anthropologist*, the flagship journal of the AAA, now carries a specially written section designed to draw more attention to the issue of public anthropology. Griffith et al. (2013: 125) set out the agenda as follows:

> Public anthropology is not a field of anthropology but a form of anthropological expression, a mechanism for connecting people like those working at Boeing, at city hall, or at Wal-Mart to work that, typically, is most often read by other anthropologists. It moves beyond the proliferation of terms (*applied, activist, feminist, engaged, critical medical, community archaeology*) to lift up the best of each, dealing with social problems and issues of interest to a broader public or to our non-academic collaborators yet still relevant to academic discourse and debate.

Although centres of traditional applied anthropology still exist, the forms of engagement anthropologists have with the worlds of policy and practice continue to diversify. During the mid 2000s some anthropologists began complaining about a lack of engagement with the 'anti-globalisation'/'anti-capitalist' movements that emerged during the 1990s, but this has changed rapidly. Factors that have contributed to this shift have included the crisis of neoliberalism in parts of Latin America, and particularly the Argentina crisis and popular insurgency of December 2001 (Schaumberg, 2008); the 2008 Western financial crisis and the growth of the Occupy movements, discussed below; and the popular street-level movements for political change that have been collectively termed the Arab spring.

As a result, the public profile of anthropology as a potential tool for understanding global issues and challenging injustices has arguably never been higher. At the 2011 AAA Annual Meeting, Gillian Tett's lecture made this very clear. Tett, an assistant editor of the *Financial Times*, trained anthropologist and author of *Fool's Gold* (2010), her account of the conditions that produced the financial crash, suggested that anthropology offered three practical strengths: the capacity to understand power, a view from the bottom up, and the potential to overcome narrow, segmented, elite forms of knowledge by 'silo busting' to achieve a more holistic perspective. Tett argued that the financial crises have brought home the idea that people matter more than technical models, showing for example that credit is a social relationship based on trust and belief among people, and not a natural or fixed state that can be assumed. To conclude, Tett suggested that anthropologists' lack of engagement and impact (relative to other social science scholars) is in part due to their excessive humility and self-criticism, in contrast, for example, with Margaret Mead's earlier far more confident engagement as a public intellectual during the 1940s.

Activist research

New thinking on what Charles Hale (2006: 97) terms 'activist research' points to new ways forward. The aim is to challenge the idea that there should be a gap between critically distanced research – cultural critique – and modes of activist engagement. He defines activist research as:

> a method through which we affirm a political alignment with an organized group of people in struggle and allow dialogue with them to shape each phase of the process, from conception of the research topic to data collection to verification and dissemination of the results. (2006: 97)

In theorising this mode of applying anthropology, Hale seeks to combine different forms of working, doing away with simplistic differences between applied and pure, involved and critical, or practising and academic. He describes the ways that many anthropologists have adopted forms of politically engaged, but non-involved, research modes. In cultural critique mode, an anthropologist's political alignment is expressed through the content of the knowledge that is produced, rather than through any direct link with people who are engaged in struggle in the real world. The contradiction here is that the anthropologist retains a primary loyalty to academia but tries to balance this with as much political commitment as is possible, but ultimately prioritises the former.

By contrast, activist research approach requires that the two loyalties become fully merged. Hale (2006: 98) suggests that proponents of activist research and cultural critique 'need each other as allies', but that this can only happen if each recognises the strengths and weaknesses of the other, and the 'underlying tensions between them'. While cultural critique may provide underpinning theoretical support for activist research, it can also create a barrier that makes it harder to progress alternative, engaged approaches. Methodologically, Hale recognises that the case for activist research can be a difficult one to make, because of the danger that such work produces forms of writing akin to '"how to" manuals, which can quickly be relegated to the marginal and devalued category of "applied anthropology"', or else take the form of tracts that seem to uncritically endorse local knowledge and politics (2006: 108). Yet he insists that, though often uncomfortable, moving between these two worlds is important both because it signifies commitment, and because it reflects the contradictions faced by the people with whose struggles anthropologists seek to ally themselves.

How might such activist research work in practice? During a conflict between Argentinian activist groups and the local municipality over local employment, pay and conditions, Schaumberg (2008: 208) describes the dilemmas he faced in balancing these two roles. In the course of his fieldwork, he was invited by activists to give his opinion at a public meeting. Despite an earlier intention to maintain distance, he did so. Having in some way now (perhaps) influenced the events he was engaged in studying, he reflected on the dilemmas that emerged:

> The UTD [Unión de Trabajadores Desocupados] would no doubt have contested the municipality's injustice without my contribution. Yet, I imagine my contribution encouraged them to stake their claim at a time when morale was very low and mobilization slow. I believe this example highlights how politically engaged fieldwork can help support justified local claims. It also shows how the public voicing of views can be not only more appropriate, but also more effective than individual discussion. (Schaumberg, 2008: 208)

In Argentina, the crisis has resulted in politicisation of local research practices in ways that may offer useful signposts for anthropologists more widely. The idea of the *investigador militante* (activist researcher) has become a familiar term, and is a model that Schaumberg suggests may be useful for constructing a more engaged anthropology in Western universities.[9]

Protest anthropology

One of the effects of the financial crisis that engulfed many Western economies in the latter part of 2008, and the unrest and protest that has followed, has been the reinvigoration of earlier forms of engaged anthropology. For example, advocates of a new 'protest anthropology' suggest that this mode implies a political and professional engagement by anthropologists that goes beyond simply being 'aligned with protest movements, revolts, and uprisings' to being 'full-fledged participants in them' (Maskovsky, 2013: 127).

Centre stage in this shift has been the Occupy movement, which took inspiration from mobilisations in Tunisia, Egypt, Spain and Wisconsin. Beginning with an initial occupation of New York City's Zuccotti Park on 17 September 2011, it quickly spread to many other cities in the US, Europe and elsewhere. Combining the occupation of public space with the attempt to build a leaderless mode of 'direct democracy' decision-making, this was a movement that expressed concern at social injustice and economic inequalities. Although the movement was criticised by the media for lacking a clear goal, it aimed to unsettle by moving beyond the formal limitations of mainstream political and civil society organisations to challenge complacency around economic crisis, social instability and environmental degradation. The ideas of anthropologist David Graeber (2013) in particular have helped underpin the movement and inform public debate through his blogs and newspaper articles, and he is credited with originating the 'we are the 99%' slogan.

Dissolving the boundaries?

The history of the applied sub-field is central to that of the wider discipline of anthropology itself. Despite the regular periods of tension that have sprung up between the 'pure' and 'applied' camps, each has ultimately contributed to the other's evolution and flourishing. At the same time, the separation between applied and mainstream anthropology has been overplayed. This has had more to do with the latter's efforts to bolster its own credibility and preserve its funding streams than with intellectual debate. If we reject this binary distinction between pure and applied anthropology, then where do we go with our exploration of anthropology and development?

The emergence of applied anthropology (along with anthropology itself) is also inextricably linked to the emergence of the concepts and practices of 'development', whether in the colonial era or later during the post-war creation of the international development industry. Indeed, James Ferguson (1996: 160) once described development as 'anthropology's evil twin'. As in the case of the pure/applied controversy, anthropological involvement in development has ebbed and flowed. This has usually been linked to the changing ideologies and approaches within the development industry (which at various times have moved closer to or further away from traditional anthropological interests), and at different times to the availability of academic jobs. For example, there were relatively low levels of anthropological engagement with development agencies during the modernisation theory era of the 1950s and 1960s, but involvement increased during the 1980s and 1990s when ideas about social development, household livelihoods and women's empowerment grew in influence among official donors. Once again, today anthropological involvement has begun to fade in relation to official mainstream development, with the current focus of the policy world on indicators, outputs and growth.

Applied anthropology is an integral component of the relationship between anthropology and development for three reasons: first, because part of the history of applied anthropology has always been bound up with the field of international development; second, because ethical dilemmas around applied engagement are still current, and have their roots in applied anthropology; and, third, because applied anthropology has always refused the distinction between 'us and them' implicit in old-fashioned colonial views of development that confine it to the 'Third World'. In subsequent chapters of this book we continue to explore the difficult issues faced by anthropologists working in and around development in the twenty-first century. Is anthropology hopelessly compromised by its involvement in mainstream development? Can anthropologists offer an effective challenge to dominant paradigms? We argue that anthropologists can suggest alternative ways of seeing and thus step outside the discourse, both by supporting resistance to development and by working within the discourse to challenge its assumptions. Most importantly, we suggest that anthropologists are in a strong position to help redefine what development should be by taking a broader view of the struggles people make to improve their lives, and drawing attention to the growing problems of inequality and the need for social justice.

3

The Anthropology of Development

When *Anthropology, Development and the Post-Modern Challenge* was published in 1996 the anthropology *of* development (ethnographic, academic analysis of both big D, little d development) was a relatively small area of study. In the intervening years it has left this somewhat marginal status far behind and become a central area within the discipline in which questions of globalisation, techniques of governance, rapid economic change, capitalism, and the everyday realities of inequality and poverty are hard to ignore. From the vantage point of the mid 1990s this is all the more remarkable, for the critiques of the deconstructivists – which we report on below – suggested that development had reached the end-game, and that all anthropologists might usefully study was its demise, or 'post-development'.[1]

In what follows we chart the emergence and establishment of the anthropology of development. In the first part of the chapter we reproduce our earlier review of the field with some minor updating work. This account, which started in the mid-twentieth century, took us up to 1996, a moment when new anthropological work on development as discourse was starting to become very influential. In the second part, we bring matters up to date, showing how the sometimes tense, occasionally hostile but often creative relationship between anthropology and big D/little d development has thrived over the last 20 years.

Anthropologists, change and development: the view from 1996

While anthropologists have long made practical contributions to planned change and policy, many have also studied development as a field of academic enquiry in itself. Although much of this work has 'applied' uses, its primary objective has been to contribute to wider theoretical debates within anthropology and development studies. As we shall see, the distinction between what Norman Long (Long and Long, 1992) calls 'knowledge for understanding' versus 'knowledge for action' is largely false and the 'anthropology of development' cannot easily be separated from 'development anthropology' (that is, applied anthropology). As Long points out, such a dichotomy obscures the inextricability of both types of knowledge, thus encouraging practitioners to view everything not written in report form as 'irrelevant' and researchers to ignore the practical implications of their findings (Long and Long, 1992: 3). Instead, the insights gleaned from knowledge produced primarily for academic purposes can have important effects upon the ways in which development is understood. This, in turn, can affect practical action and policy. Indeed, rather than necessarily being trapped within the dominant discourses of development, we suggest that the anthropology of development can be used to challenge its key assumptions and representations, both working within it towards constructive change, and providing alternative ways of seeing which question the very foundations of developmental thought. Research which focuses upon local resistance to development activities, or which contradicts static and dualistic notions of traditional and modern domains, are just two examples.

Since no society is static, change should be inherent in all anthropological analysis. However, this has not always been the case. While in its earliest phases the discipline was based upon models of evolutionary change, from the 1920s until the 1950s British social anthropology was dominated by the functionalist paradigms of Malinowski and Radcliffe-Brown (Grimshaw and Hart, 1993: 14–29). These presented the 'exotic' peoples studied as isolated and self-sufficient; social institutions were functionally integrated and each contributed in different ways to social reproduction. Rather than continually changing according to wider political or economic circumstances, such societies were therefore presented in ahistorical terms, functionally bound together by the sum of their customs and social institutions.

By the 1960s and early 1970s, structural-functionalism was increasingly superseded by the structuralism of Lévi-Strauss.[2] While based on quite different theoretical premises from those of structural-functionalism, this too was largely uninterested in change, seeking out the logics of thought and social practice in binary oppositions which, the structuralists argued, underlie all human culture. Although structural-functionalism and structuralism were not the only paradigms in anthropology over these periods, and writers such as Leach challenged the static nature of structuralist accounts,[3] in general, history and economic change were not given much consideration by the mainstream. Indeed, cultural units were often portrayed in ethnography as isolates; if the forces of market or state were mentioned, they were presented as autonomous forces, impinging from the outside (Marcus and Fischer, 1986: 77).

In spite of these trends, individual anthropologists have long been studying the effects of economic change, development projects and global capitalism. Within some branches of anthropology, such work has always been closely connected to theory: French Marxist anthropology is just one example (see Bloch, 1983). Meanwhile recognition of the historical embeddedness of ethnography started to grow from the 1970s. This was associated with anthropology's bout of self-criticism and reflexivity, and with wider critiques of the way in which Western scholarship presented timeless, ahistorical 'others' (Marcus and Fischer, 1986: 78). By the early 1990s, understanding cultural and social organisation as dynamic, rather than fixed or determined by 'set' essentials had become central to anthropology, and it was widely appreciated that culture does not exist in a vacuum but is determined by, and in turn determines, historically specific political and economic contexts.

In the following review we cannot begin to discuss the vast range of anthropological work which places change at the centre of the analysis. Even if we only included research which focuses directly on situations where capitalist forms of production, exchange or labour relations have recently been introduced, the potential range of material is huge. It is not our intention to produce a comprehensive survey of such work, nor do we intend to discuss the many non-anthropological studies of development. Instead, in what follows we provide a quick 'taste' of the ways in which anthropologists have tackled economic change and growth, whether this was deliberately planned or more spontaneous. As we shall see, while not all of this work explicitly questions or challenges the dominant development discourse, some of it does so implicitly.

In general, the anthropology of development (and by this we mean planned and unplanned social and economic change) can be loosely arranged around the following themes:

1. The social and cultural effects of economic change.
2. The social and cultural effects of development projects (and why they fail).
3. The internal workings and discourses of the 'aid industry'.

Some work covers all these themes; the first two, in particular, are closely interrelated. Clearly too, the potential applicability of the different analyses varies. Work which addresses the second issue, for example, often aims to affect policy as well as add to academic debate. It is generally sympathetic rather than completely condemnatory of development practice, assuming that the understandings which it provides are crucial tools in the struggle to improve development from within. In this sense it tends to blur the boundaries between academic and applied anthropology. In contrast, anthropologists interested in the last question are usually less interested in aiding development practitioners; while their insights may have policy implications, such work rarely ends with practical recommendations. Instead they hope to problematise the very nature of development. As we shall see, the three themes can also be linked, albeit very loosely, to historical changes within both development and anthropology.

The social and cultural effects of economic change

Although the study of economic change has not always been academically fashionable, individual anthropologists have long been grappling with it. As we saw in the last chapter, the relationship between anthropology, its practical application and questions of change were originally (in British social anthropology at least) entangled with colonial rule, especially in Africa. Malinowski was the first anthropologist to propose a new branch of the subject: 'the anthropology of the changing native' (1929: 36; also cited in Grillo, 1985: 9), sending students such as Lucy Mair to Africa to study social change rather than more abstract theoretical principles. Even Evans-Pritchard – accused today of having remained silent in his famous ethnographic writings on the Nuer about the frequent aerial bombings of their herds as part of the colonial government's 'pacification' programme in the 1930s during his fieldwork – had argued in earlier work that the Nuer were in a state of transition, their clans and lineages broken up by endless wars (discussed by Kuper, 1983: 94). Let us start, then, with some of the early work of British anthropologists working in colonial Africa.

Rural-to-urban migration and 'detribalisation'

One of the earliest collective efforts to make sense of economic and political change in Africa was embodied by the Rhodes-Livingstone Institute in 1937. While it was originally assumed that the institute's research would concentrate upon 'traditional' African rural life, the director, Godfrey Wilson, made it clear that he was most interested in urbanisation and its influence on rural life (Hannerz, 1980: 123). In the books which resulted from Wilson's research in Broken Hill (now Zambia) (Wilson, 1941, 1942), he argued that while central African society was normally in a state of equilibrium, destabilising changes had been introduced which had led to disequilibrium. These changes were mostly the result of the increasing influence of capitalist production within the region: industrialisation, and growing rural-to-urban migration. As in Zambia's Copperbelt, Broken Hill was dominated by the European mining industry, which largely determined African migration to and settlement within it. Because colonial policy discouraged permanent settlement, most of the male migrants working for the mines moved between their villages and the town. Wilson suggested that destabilisation might be offset if this policy were reversed and proposed that eventually the changes would be incorporated by the social system, leading once more to equilibrium.

Urban migration in Rhodesia, as in other parts of Africa, had a dramatic effect on rural areas. Many villages lost a large proportion of their male labour force, and most migrants could not afford to send back enough remittances to compensate. The work of other anthropologists confirmed this gloomy view of labour migration, linking it with decreasing agricultural output (A. Richards, 1939) and cultural decay (Schapera, 1947). While this perspective was to change in later studies, which suggested that rural-to-urban migration in Africa might be a force of modernisation (for a review, see Eades, 1987: 3), other, more contemporary work has taken up similar themes. Colin Murray's analysis of labour migration in Lesotho, for example, shows how rural life has been deeply structured by its dependence on the export of labour to South Africa. Oscillating male migration has generated economic insecurity, marital disharmony and the destruction of traditional kinship relations. In other words, capital accumulated at the urban core takes place at the expense of the rural periphery (Murray, 1981).

While this body of work raised questions about the relationship of societies on the 'periphery' to the global political economy, research based in the Copperbelt towns greatly contributed to anthropological understanding of ethnicity. The Rhodes-Livingstone Institute, and the continuation of its work under Max Gluckman at the University of Manchester, focused largely upon social and cultural forms within the mining towns. Central to much of this work was the issue of 'detribalisation'. This was the argument that once individuals moved to the towns their tribal bonds became less important, being superseded by class or workplace affiliations. Research showed that this was not necessarily the case. Rather, tribal identities and obligations changed, and were used in different ways in the urban setting. Mitchell's seminal analysis of *The Kalela Dance* (1956), Epstein's *Politics in an Urban African Community* (1958) and Cohen's (1969) slightly later analysis of Yoruba traders and the use of ethnicity for political and economic interests raised questions of identity, ethnic conflict and cultural diversity, which are of central interest to anthropologists today.

Agricultural change: polarisation

While the anthropology of urbanisation in Africa during this period was rooted in pre-war colonial policy, studies of rural change in South and South East Asia were largely influenced by postcolonial states' efforts to modernise in the 1950s and 1960s. Much of this work indicated that the transition to cash-cropping, mechanisation and the growing importance of wage labour had a range of social effects, not least of which was increasing polarisation and the proletarianisation of the rural poor. It seemed that the 'Green Revolution' and other modernisation strategies were unlikely – at least for the foreseeable future – to diminish poverty. These critiques contributed to growing scepticism about the 'trickle-down' effects of economic growth, and added to calls for a shift in policy towards 'basic needs' and the targeting of particularly vulnerable groups.

Let us start with Clifford Geertz's account of Indonesian agricultural change, *Agricultural Involution* (1963). By providing an historical account of Indonesian agriculture, Geertz showed how colonial policies encouraged the development of a partial cash economy in which peasant farmers were forced to pay taxes to support plantation production for export. This, alongside the policies of the post-independence elite, contributed to growing dualism. The majority of farmers formed a labour-intensive sector in which they were unable to accumulate capital and produced mainly for subsistence, while another sector became capital-intensive and techno-logically advanced under colonial management. Economic stagnation in Indonesia was seen as deeply structured not only by history and ecology but also by social and cultural factors (Geertz, 1963: 154).

In contrast to Geertz's adventurous multidisciplinary approach, other anthropologists, in a more traditional mode, focused upon the effects of economic change at the micro level. In South Asia, two of the most famous of these are Bailey's *Caste and the Economic Frontier* (1958) and Scarlett Epstein's *Economic Development and Social Change in South India* (1962). In a later work, *South India: Yesterday, Today and Tomorrow* (1973), Epstein discusses the effects of the introduction of new irrigation techniques and the growing importance of cash-cropping to two villages in South India. In the village of Wangala, where farmers were increasingly producing for and profiting from a local sugar refinery, the changes had not led to major social readjustment. The village continued to have few links with the external economy and the social structure remained largely unaltered, due to both the flexibility of the local political system and the fact that the economy was still wholly based upon agriculture. In contrast, in the second village, Dalena, which had remained a dry land enclave in the midst of an irrigated belt, male farmers were encouraged to move away from their relatively unprofitable agricultural pursuits and participate in other ways in the burgeoning economy which surrounded them. Some became traders, or worked in white-collar jobs in the local town. These multiple economic changes led to the breakdown of the hereditary political, social and ritual obligations, the changing status of local caste groups and the rise of new forms of hierarchy.

The different changes in each community indicated that processes of capitalist transformation were far from homogeneous, even within the same region. Instead, economic and technological changes interrelated with pre-existing social and cultural forms in a variety of ways, and have diverse consequences. Epstein's work also showed that in both villages social differentiation was increasing. In Wangala, despite the government's abolition of 'untouchability' in 1949, those lowest in the caste hierarchy remained in the same position. The gap between the poorest and the richest was, however, growing. Likewise, traditional bonds between employers and labourers were largely intact. In Dalena there had been some compromises over 'untouchability', but, at the same time, the security of labourers had diminished; the poorest were becoming increasingly temporary and wholly dependent upon their small wages rather than the traditional patronage of their employers.

A wide literature supports Epstein's view that the modernisation of agriculture (the introduction of new technologies, cash-cropping, wage labour) in South Asia has contributed to growing rural polarisation. Much of this constitutes a critique of the Green Revolution, correcting initial claims that the 'package' of agricultural innovations would cure all hunger. Again, the effects of the innovations depend partly upon pre-existing social relations. John Harriss's study of social changes in North Arcot, South India, for example, showed that while farmers were increasingly linked to external markets and government institutions, traditional patron clientelism is reinforced (J. Harriss, 1977). Meanwhile, the poorest were worse off, for alongside the new technology came increasing competition over scarce resources, together in some cases with displacement of labour by new technology (Farmer, 1977). These effects, added to the non-adoption of many parts of the package, were noted across the world (Pearse, 1980).

Modernisation was thus not nearly as simple as many theorists during the 1950s and 1960s had assumed. While writers such as Epstein were not engaged in the critical deconstruction of 'development' which was to emerge several decades later, their ethnography vividly demonstrated the flaws in the conventional developmental thinking of the time. They also contributed to wider debates within anthropology; for example, Bailey and Epstein were just two of many anthropologists working in South Asia on the changing nature of caste and kinship institutions during this period (see also Vatuk, 1972; Breman, 1974). More recently, a growing body of work examines the varied effects of agrarian change and industrialisation in India, showing how new forms of inequality arise from, or co-exist with those of earlier periods (for example, Bremen, 1999, 2004; Munster and Strümpell, 2014; Carswell and De Neve, 2014; Corbridge and Shah, 2013).

Capitalism and the 'world system'

As notions of modernisation and 'trickle-down' effects were being increasingly questioned by both anthropological findings and the evident failure of many development policies, other researchers were turning their attention to the relationship of local communities and cultures to the global political economy. This can be linked to the growing dominance during the 1970s of theories of dependency, and especially to Wallerstein's world system theory (Wallerstein, 1974), as well as the use of Marxism in the 1970s and 1980s by some anthropologists (Bloch, 1983). Rather than analysing development in terms of the transformation of otherwise untouched or 'traditional' communities by economic or technological innovations, the emphasis here was more upon the ways in which societies on the 'periphery' had long been integrated into capitalism, and thus on the social and cultural expressions of economic and political dependency, interaction (articulation) and/or resistance. Such work placed indigenous experiences and expressions of history at the centre of the analysis; and colonialism and neo-colonialism are often key to this.[4] It is worth noting that much of this research was carried out in Latin America, where dependency theory originated. Like the questions raised by dependency theory, those arising from this approach are less easily translated into national or regional policy. It critiqued the basis of development discourse, rather than attempting to work within it.

Eric Wolf's *Europe and the People without History* (1982) remains a classic attempt to fuse neo-Marxist political economy with anthropological perspectives. This was an ambitious attempt to place the history of the world's peoples within the context of global capitalism, showing how the history of capitalism has tied even the most apparently remote areas and social groups into the system. In it, Wolf argued that Marxist concepts such as the mode of production involved social and cultural, as well as technical, aspects. Since he concentrates on the macro level, however, his analysis of culture is rather limited (Marcus and Fischer, 1986: 85). As others have pointed out too, the spread of European capitalism is far from being the only history to be told of the 'people without history' (Asad, 1987). Similar themes are taken up in Worsley's *The Three Worlds: Culture and World Development* (1984), which provided further analysis of the relationship between local cultural expressions and the exploitative workings of global capitalism.

The integration of political economy and history into ethnographic analysis opened important doors in anthropology during the 1980s, contributing to some of the most exciting work to be produced in recent decades. In this, the mediation between structure and experienced practice was central, indicating the diverse ways in which people struggle to construct meaning and act upon the forces which often subjugate and engulf them. Comaroff's *Body of Power, Spirit of Resistance* (1985), an analysis of the interrelationship between history and culture among the Baralong boo Ratshidi, a people on the margins of the South African state, was a classic example of such an approach. David Lan's *Guns and Rain* (1985), an ethnography of rural revolution in Zimbabwe, and the place of guardians of rain shrines in it was another example.

Drawing more directly from neo-Marxist theories of dependency, two important studies by anthropologists working in Latin America indicated both the extent to which groups are linked into global capitalism, and the ways in which this is interpreted and culturally resisted. Michael Taussig's *The Devil and Commodity Fetishism in South America* (1980) offered an account of the cultural as well as economic integration of Colombian peasants and of Bolivian tin miners into the money economy and proletarian wage labour. The Colombian peasants who seasonally sell their labour to plantations present the plantation economy and profits made from it as tied to the capitalist system, and thus to the devil. Plantations are conceptualised as quite separate from the peasants' own land; in the former, profit-making requires deals to be made with the devil, whereas in the latter it does not. In the Bolivian tin mines, workers worship Tio (the devil), who Taussig argues is a spiritual embodiment of capitalism and a way of mediating pre-capitalist beliefs with the introduction of wage labour and industrialisation. Similar themes are explored in June Nash's *We Eat the Mines and the Mines Eat Us* (1979). Again drawing on Latin American dependency theory and on Marxist analysis of ideology and class consciousness, Nash explored the cultural and social meanings given to capitalist exploitation at the periphery.

Taussig's and Nash's work concentrates largely upon local ideologies of capitalist integration without directly questioning models of dependency and global exploitation. Other anthropologists, however, have added to the growing critique of dependency theory and its eventual fall from grace during the 1980s. In Norman Long's research in the Mantaro Valley of central Peru, for example, he found that neo-Marxism only offered limited insights (Long, 1977). Instead, his findings challenged dependency theorists' assumptions that integration into global capitalism could only lead to stagnation on the periphery. In his research he found both growth and diversification in the Mantaro Valley. Indeed, some groups had been highly entrepreneurial, generating considerable small-scale capital accumulation. Local producers had also developed a complex system of economic linkages, which was far from simply determined by the 'centre'. Contrary to the assumptions made by dependency theory, there were no obvious chains of hierarchy linking them to the metropolis, or to the mining corporation. Through anthropological methods (interviews, situational analysis, life history studies, social network analysis and so on), Long's research allowed him to indicate the different responses to change of the actors themselves, revealing a far more complex and dynamic situation than structural Marxist analysis of the macro level could ever allow.

Most important, perhaps, was Norman Long's use of the notion of human agency; the recognition that people actively engage in shaping their own worlds, rather than their actions being wholly pre-ordained by capital or the intervention of the state (Long and Long, 1992: 33).

Similar conclusions had been reached by researchers working in squatter settlements in Latin America. Prompted in part by the findings of Mangin (1967), a sociologist, and Turner (1969), an architect, various writers argued during the 1960s and 1970s that rather than being 'slums of despair' the settlements were in fact 'slums of hope' (Lloyd, 1979). Invasions of land were carefully planned and people worked together to obtain water, electricity and roads for their settlements, forming committees and gaining a voice through electing local politicians to state and metropolitan bodies. Rather than being passive 'victims' of international and national structures of exploitation, the squatters were active agents, working hard to transform their economic and social standing. Whether or not they were always successful depended to a large degree upon state policies towards squatting. They were not, however, 'marginal'; instead, they were marginalised by wider contexts, even while striving to improve themselves (Perlman, 1976).

While stress on the perspectives of actors, rather than the 'systems' of which they are a part, has always been central to anthropology, such ideas were widely taken up within development studies in the 1980s and 1990s, partly perhaps both because they pointed to constructive changes which can be made in policy, and because the 'developmental' message is essentially optimistic: people are not wholly constrained by exploitative superstructures or the 'world system'; they are active agents and, if there is to be intervention, merely need to be 'helped to help themselves'. During the 1980s growing emphasis was put upon the subjects of development projects as 'actors', adding to ideas about participatory development, the 'farmer first' movement and the importance of 'indigenous knowledge', all of which will be discussed in later chapters. For now, however, let us turn to another major contribution of anthropology to the understanding of social and economic change: the analysis of gender relations.

The gendered effects of economic change

Alongside the first stirrings of feminist anthropology in the early 1970s came the growing recognition that economic development has differing effects on men and women. Increasing interest in the relationship between gender and development was precipitated largely by the publication of Ester Boserup's ground-breaking *Woman's Role in Economic Development* (1970). In this, Boserup pointed out that the sexual division of labour varies throughout the world and that, contrary to Western stereotypes, women often play a central role in economic production. Nowhere is this truer than in African contexts, which Boserup contrasts with 'plough economies' where, she asserts, women are secluded and play a diminished role in production (an assumption which in fact is largely unfounded). Women's varied productive roles, she argues, are due to population pressure, land tenure and technology. As economies become more technologically developed, women are increasingly withdrawn from production or forced into the subsistence sector, while men take centre stage in the production of cash crops. These changes are not automatic, but have been influenced by ethnocentric colonial policies which assumed that women were not involved in agricultural production and thus bypassed female farmers in favour of men.

Boserup's work was an important catalyst for an enormous literature on the effects of development on gender relations. Much of this focuses on particular projects and policies, which we shall discuss in the next section of this chapter. Other researchers looked at the wider relationship between capitalist change and gender. This was not a new debate: as early as 1884 Engels had discussed the relationship between the subordination of women and the development of class relations alongside the privatisation of property, in *The Origin of the Family: Private Property and the State* (1972). While lying largely dormant in anthropology up until the 1960s, such concepts were eagerly taken up and reworked by a new generation of feminist anthropologists during the 1970s (for example, Leacock, 1972; Sacks, 1975). While not all academics working on what became known as 'GAD' (gender and development) were anthropologists, much of their work drew heavily on the field of feminist anthropology, which during the 1970s was growing in intellectual credibility and theoretical rigour (for a summary, see Moore, 1988). Not all of this work was directly concerned with economic 'development'; some feminist anthropology, for example, involved the re-study of the subjects of ethnographic classics from a feminist perspective,[5] while other work focused on women's supposed universal subordination and its cultural expressions.[6]

The capitalist transformation of subsistence economies is generally acknowledged as having a negative effect on women (see for example Afshar, 1991). Change in land tenure, labour migration and a growing market in land and labour have all contributed to the marginalisation of women from processes of change, relegating women to subsistence production. The 'feminisation of subsistence' thesis is explained in two ways (Moore, 1988: 75). First, since women have reproductive as well as productive duties (they must feed, clothe, shelter and emotionally support their families), they are less free to spend time producing cash crops. Thus while men may be able to experiment with new technologies and production for exchange, women must first and foremost produce the subsistence foods on which their households depend. Second, male labour migration leaves women behind to carry the burden of supporting the subsistence sector.

While the 'feminisation of subsistence thesis' is in many ways problematic (for example, in many parts of Asia men still play a dominant role in subsistence agriculture), it raised similar issues to those of research on the Green Revolution: economic change has differential social effects. But rather than these differential effects being experienced *between* households, feminist anthropology indicates that they exist *within* them. Equality cannot be taken for granted at any level of social organisation (Folbre, 1986).

Ann Whitehead's research on the Kusasi in Ghana is an excellent example of these points, demonstrating that we need to deconstruct concepts of both the household and the sexual division of labour, which involves not just different tasks but also different access to resources (Whitehead, 1981). Among the Kusasi there are two types of farm, private and household, and men and women have different access to resources, which they do not pool. The main constraint on productivity is access to labour rather than to land. Productivity depends to a large extent on the degree to which social networks – and thus labour – can be mobilised. Men are better able to do this than women: while they can call upon the labour of their wives, women can only use male household labour by paying for it with drink and food. Meanwhile men are often able to commandeer community-wide work parties. As this and other research clearly indicates, projects aimed at increasing productivity thus often have to negotiate complex economic and social relations which are embedded in the local cultural context. Assumptions cannot be made about the nature of households, the distribution of resources within them, or the social relations of production.

The work of feminist anthropologists in analysing the gendered effects of economic change has made a substantial contribution both to development studies and to anthropology. We shall discuss the former in the next section. Within academic anthropology, during the 1970s and 1980s feminists pushed a new domain of study onto the anthropological agenda: the cultural, political and economic construction of relations between men and women. This involved radically unpicking various anthropological concepts which had formerly been treated as unproblematic: the household, the 'domestic mode of production' and the division of labour were all deconstructed and reconstituted in far more incisive terms (see, for example, O. Harris, 1981). Feminist anthropology also sounded the final death knell for structural-functionalism: given what it told us about power, resistance and the cultural hegemony of patriarchy, the notion that societies are functionally integrated and in equilibrium was clearly no longer credible. The pressure from feminist anthropology to deconstruct androcentric categories and assumptions can also be seen as the precursor to the increasingly reflexive nature of anthropology in the 1980s and into the 1990s.

The social and cultural effects of development projects (and why they fail)

Many of the texts discussed above have been concerned with the issue of social and cultural impacts. Here, however, we shall consider work which focuses specifically upon development projects, a discussion that continues one we began in Chapter 2. Rather than treating them as external forces which affect the social group or community being studied, this may involve studying the internal workings of the projects themselves, an issue we shall return to in the next section. Much (but not all) of this work is largely sympathetic to the developmental effort (Ferguson, 1990: 9), presenting it as a collective effort to fight poverty, rather than a form of imperialism or dependency. The research agenda thus tends to be dominated by pragmatic assessments of what goes wrong with development projects, and how they could be improved. Within the anthropology of development, this body of work is thus the most easy to apply practically, and texts often end with lists of concrete recommendations. As we shall see, anthropologists have tended to call for the same solutions: local participation, awareness of social and cultural complexities, and the use of ethnographic knowledge at the planning stage.

One of the most common criticisms made by anthropologists of development planning has always been that it is done in a 'top-down' manner: plans are made by distant officials who have little idea what the conditions, capabilities or needs are in the area or community which has been earmarked for developmental interventions. By imposing such plans on people, rather than allowing them to participate in the decision-making process, it is argued, interventions are doomed to failure, for development can only ever be sustainable if it is from the 'grassroots'. Criticisms are thus aimed not at development per se, but at the way in which it is carried out. Changes in policy and practice, it is optimistically assumed, will mean that development projects are increasingly successful in helping the poor.

Robert Chambers's *Rural Development: Putting the Last First* (1983) was a seminal statement of this position and drew heavily upon the insights of anthropology. In this and subsequent publications, Chambers attacked the biased preconceptions of development planners, most of whom have only a very shaky understanding of rural life in so-called developing societies (Chambers, 1983, 1993). Their urban bias, the use of misinformed research and statistics, and their neglect of local solutions and knowledge means that development policies and projects can never succeed, for they do not understand the hidden nature of rural poverty. The only solution, Chambers argues, is to 'put the poor first' and, most importantly, enable them to participate in projects of their own design and appraisal.

Tony Barnett's *The Gezira Scheme: An Illusion of Development* (1977) was a classic critique of 'top-down' development. The Gezira Scheme was a colonial economic development project begun during the 1920s which was intended to introduce intensive irrigated cotton production in Sudan. Despite the apparent wellbeing of the Gezira peasants, Barnett suggested that the project led to stagnation and dependency. The scheme was huge, involving 12 per cent of the total cultivated area in Sudan and the leasing of government land to over 80,000 tenants. These cultivated cotton for export, and were allowed neither to have more land than they could cultivate, nor to sell it. Barnett argued that the relationship between the cultivators and the Sudan Gezira board was paternalistic and authoritarian, based on British efforts to control 'black' labour. This meant that cultivators had few incentives to be innovative, and Sudan remained largely dependent upon foreign markets for its cotton. In such a context, aid is more to do with 'neo-colonialism' than even attempting to help the poor. To this extent Barnett's work has theoretically more in common with neo-Marxist analyses of the role of aid in reproducing the dependency of the periphery than with the more positive approach of writers such as Chambers.

Most anthropological critiques of development projects criticise planning which is insensitive to the cultural and social complexity of local conditions and thus to the diverse effects of externally induced change. Let us turn to work which examines the effects of this on gender relations within development projects. As we have seen, anthropological research has had a major impact on understandings of the effect of economic change on gender relations. Not only have feminist anthropologists provided ethnographic accounts of this, they have also developed various analytic tools (the division of labour, production and reproduction, the household) to illuminate why development tends to have such different effects on men and women. Much of this work focuses on the effects of specific development projects. There is a vast literature on this; here, we intend only to give a brief introduction to some of the main issues and texts.

By misunderstanding the sexual division of labour, access to resources in the household and women's double burden of productive and reproductive work, development planning and projects frequently lead to the marginalisation of women. This is because of both pre-existing gender relations (which mean that men are better placed to appropriate new economic opportunities) and the patriarchal assumptions of planners. This process began with colonial administrators, who imported ethnocentric notions of 'the place of women', and continues today through the work of Western development planners. In *The Domestication of Women* (1980), Barbara Rogers argued that Western development planners make a range of Western, and thus patriarchal, assumptions about gender relations in developing countries. It is often assumed, for example, that farmers are male, that women do not do heavy productive work and that nuclear families are the norm. Through androcentric and biased research, such as the use of national accounting procedures and surveys which assume that men are household heads, women become invisible. Women are thus systematically discriminated against, not least because there is discrimination within the development agencies themselves. Again, this process began with the 'men's club' (Rogers, 1980: 48) of colonial administration, but is continued today in organisations such as the Food and Agricultural Organization and the World Bank.

The answer, Rogers argued, was not simply more projects for women, for these often produce a 'new segregation', in which women are simply trained in domestic science or given sewing machines for income generation. Instead, gender awareness must be built into planning procedures, a process which will necessarily involve reform of the development institutions involved. Similar conclusions were drawn by other, policy-oriented writers, such as Staudt (1990, 1991) and the contributors to *Gender and Development: A Practical Guide* (Ostergaard, 1992).

While Rogers took a more general view of the discriminatory effects of planned development, other writers concentrate on particular projects.

Dey's (1981) account of irrigation projects in the Gambia showed that by assuming that men controlled land, labour and income, the projects failed to increase national rice production and increased women's dependency on men. Within the farming system of the Mandinka, crop production is traditionally dominated by collective production for household consumption (*maruo*), but also involves separate cultivation by men and women on land they are allocated by the household head in return for their *maruo* labour (*kamanyango*). Crops from this land are the property of the male or female cultivators. However, under rice irrigation projects sponsored by Taiwan (1966–74), the World Bank (1973–76) and China (1975–79), only men were given *kamanyango* rights to irrigated land; in other irrigated plots designated as *maruo*, men increasingly used women's skilled collective labour, but were able to pay them low wages because of the lack of other income-generating opportunities available to women. Women's traditional economic rights were thus systematically undermined by the projects, a process which had started during the colonial period, when once more the reciprocal rights and duties of farming were undermined by policies which encouraged male farmers to produce cash crops and failed to recognise the central role of female producers. By ignoring the complexities of the farming system and concentrating on male farmers, the projects thus not only disadvantaged women, but lost out on their valuable expertise.

Because gender relations are culturally specific, development projects have different effects according to where they are carried out and the ways in which they are implemented. Data from Asia, for example, showed that whereas farm mechanisation led to declining female labour in rice-farming villages in the Philippines, in Japan female participation has remained relatively high (Ng, 1991: 188). In her case study of the introduction of advanced mechanisation in a rice-growing village in West Malaysia, Ng (1991) showed how women's participation in the labour force had declined. The Northwest Selangor Integrated Agricultural Development Project, launched in 1978, aimed to increase yields, maximise income and thus alleviate rural poverty by the introduction of Green Revolution-type technologies. While this has indeed led to higher yields, the division of labour by gender has been transformed, significantly reducing women's contribution to farming and thus leading to a reduction in their productive skills. With their displacement from rice production, their domestic role is increasingly important to women, due to the prevailing gender ideology which places priority on women's reproductive work; this is encouraged by both the state and rural patriarchy. Class is an important factor too. While women from rich and middle-income households have increasingly (and apparently happily) retreated to the domestic arena, women from poor households need to work to raise the cash for the new inputs necessary for increased productivity. There are thus two broad trends: patriarchal households among the rich and middle-income households, and female-headed households among the poor.

The analyses of development projects by feminist anthropologists have had important implications for policy-makers.[7] There is not the space here for a comprehensive review of the effects of women in development (WID) and GAD on development policy.[8] Suffice it to say that since 1975, with the start of the first UN Development Decade for Women, gender was increasingly acknowledged as a central issue within development circles. Many agencies developed explicit policies on gender, employing 'experts' to ensure that their projects give sufficient consideration to the interests of women. The World Bank, for example, created a WID unit, while UNIFEM (United Nations Development Fund for Women) has been a United Nations agency since 1985 (Madeley, 1991: 29). Gender training also took off from the 1980s, with agencies funding the training of both their own staff and staff in local government and other institutions in recipient countries.[9] Whether or not these efforts have had any real impact on improving the detrimental effects of development on women is, however, debatable. Indeed, some argued that WID policies and training reproduce ethnocentric assumptions about the nature of gender and women's subordination; that they co-opt radical feminist critiques into the development discourse, thus neutralising them. We shall return to these issues later.

Closely related to anthropological critiques of 'top-down' planning is the criticism that planners fail to acknowledge adequately the importance, and potential, of local knowledge. Instead, projects often involve the assumption that Western or urban knowledge is superior to the knowledge of the people 'to be developed'; they are regarded as ignorant although, as anthropologists have repeatedly shown, they have their own areas of appropriate expertise. This is tied to the 'farmer first' movement (Chambers et al., 1989; Scoones and Thompson, 2009). It also raised interesting questions about the interrelationship of different forms of knowledge, which we shall return to in the next section. For now, however, let us consider cases where 'top-down' planning means that not enough is known about the culture or conditions of an area or target group before a project is embarked upon.

Development projects often fail because of the ignorance of planners rather than the ignorance of the beneficiaries. This might involve a range of factors, such as local ecological conditions, the availability of particular resources, physical and climatic conditions and so on. The result is inappropriate intervention, which may end in disaster. (An example is the infamous Groundnut Scheme in Tanzania described earlier; see Wood, 1950.) The success of all projects depends upon whether or not they are socially and culturally appropriate, yet, ironically, it is these factors which tend to be least considered. Much literature therefore focuses upon the need for ethnographic knowledge at the planning stage of project design (for example, Mair, 1984; Hill, 1986; Pottier, 1993). Again, this perspective is ultimately optimistic: with better planning (and the use of ethnography), it is assumed, development projects will succeed in helping poor people.

Mamdani's classic analysis of the failure of the Khanna study, an attempt to introduce birth control to the Indian village of Manupur, was a fascinating account of developmental top-downism and ignorance (Mamdani, 1972). Because of the cultural and economic value of having as many children as possible, Mamdani argued that population programmes were unlikely to have much success in rural India. Programme planners in the Khanna study assumed that villagers' rejection of contraception was due to 'ignorance', thus completely ignoring the social and economic realities of the village. Once again, anthropological methods and questions, rather than bureaucratic planning, reveal the true constraints on 'successful' development. While Mamdani is to be congratulated for powerfully illustrating the cultural and economic influences on family planning uptake, he can also be criticised for assuming that local attitudes to family planning are homogeneous. Other work questioned this, indicating that men and women often have very different views and that it is men who usually control eventual fertility decisions. This is an area where feminist researchers clearly have much to contribute (for further discussion, see Kabeer, 1994: 187–222).

Pottier's edited collection, *Practising Development*, took these issues substantially further. It also clearly reflected changes within developmental practice, wherein notions of participation and 'farmer first' began gaining increasing currency (Pottier, 1993). While all contributions took for granted the need for anthropological insights at the planning stage and show how this is already a common practice for some organisations – for example, the International Fund for Agricultural Development (IFAD) (Seddon, 1993) and Band Aid (Garber and Jenden, 1993) – most examined how social science perspectives could be effectively incorporated into development programmes. This was not simply a matter of becoming literate in the local culture, as if it were composed of essential and accessible elements. A critical perspective here is that 'the social worlds within which development efforts take shape are essentially fluid' (Pottier, 1993: 7). Gatter's Zambian case study, for example, demonstrated how farming practices tend to be systematised by development workers, who thus misunderstand their complexity and fluidity (Gatter, 1993). To avoid such misrepresentations, and make ethnographic knowledge meaningful, there must therefore be a continual collection of ethnographic data. This research need not necessarily be carried out by expatriate consultants but can be done by trained field staff, especially those in NGOs. Crucially, Pottier's collection adopts an approach flourishing in the anthropology of development: that of studying development bureaucracies and institutions in themselves, as well as the discourses which they produce. Let us turn to our third theme.

The internal workings and discourses of the 'aid industry'

Rather than simply viewing development as an external force, which acts upon the 'real' subjects of anthropological enquiry (the 'people'), anthropological accounts of development have come to treat its institutions, political processes and ideologies as valid sites of ethnographic enquiry in themselves. While this approach is not solely confined to the late 1980s and 1990s, its increasing dominance reflects contemporary trends in anthropology. Before turning to this, let us start with the anthropology of development planning.

Anthropologists have been aware of the need to study the internal working of development institutions for some time, although studies of administration are usually focused far more on the recipients of planned change than on the 'developers'. As we saw in Chapter 2, early work in the applied anthropology tradition such as H.G. Barnett's *Anthropology in Administration* (1956) dealt mainly with the practical uses to which anthropological knowledge could be put by administrators, using examples drawn from the author's experience of working in the Trust Territory of the Pacific Islands, and only occasionally turns its gaze upon the system itself. Cochrane's *Development Anthropology* (1971) emphasised the need for administrators, under the guidance of anthropologists, to recognise the cultural issues surrounding development in addition to the more familiar economic and technological aspects in which they are trained. Belshaw's *The Sorcerer's Apprentice* (1976) sought to draw anthropological concerns away from the 'exotic' towards real policy issues in the dominant culture and to counter the tendencies of administrators only to 'know and control'.

More ambitiously, Robertson's *People and the State* (1984) attempted to analyse planned development as a political encounter between the people and the state. Development agencies, he argued, are premised on the need to turn an unreliable citizenry into a structured public; development interventions are thus the site of contest between the people and bureaucracy (1984: 4). Much of the book recounts the history of planning, from post-revolutionary Russia and colonial planning to the economic planning of contemporary Third World states. Robertson also makes a plea for anthropology to become more centrally involved in development. Although historically anthropology has been weak on state theory, he suggests that it can potentially offer an overview of the whole planning process, thus making a vital contribution to wider understandings of development. Like Cochrane, Robertson was interested in the practical uses of anthropology and appears to be optimistic about the potential of planned change. As he concludes: 'anthropology may ultimately prove its worth by helping to explain a confused and lethally divided world to itself, and to indicate humane and realistic prospects for progress' (Robertson, 1984: 306).

Project and planning ethnography is linked to shifting paradigms within development studies. Here too, there was increasing recognition that the realities within which people act and make decisions are multiple and changing. This was closely related to the interest in actor-oriented research, in which the worldviews of individual actors (rather than passive target groups or beneficiaries), and the interfaces between them and bureaucratic institutions, are the focus of study (Long and Long, 1992). Notions of 'farmer first' development, and participation, were particularly influential here. On a slightly different level, recognition of the need to understand (and then change) the workings of bureaucracy (in, for example, writings on gender and development: Staudt, 1990, 1991) were also important.

The authors discussed above present planning as a relevant and important area of anthropological research. All share – in different degrees – a practical agenda: to improve the planning process, usually with help from anthropological inputs. In contrast to this, other work deconstructs and problematises the very notion of development by analysing it as a form of discourse. This work is not intended to be instrumental for policy-makers, as it critiques the epistemological assumptions within which they work. Instead, it has far-reaching implications for the way in which 'development' is conceptualised, pointing to a radical reappraisal of the ways in which global poverty and inequality are conceptualised and tackled. As we shall see, such work has been strongly influenced by postmodern understanding of culture as negotiated, contested and processual. Social realities in these accounts are multiple, and change according to context. To this extent writers do not search for objective 'truths' about development or its effects, but seek to understand the ways in which it is socially constructed and in turn constructs its subjects. Much of this has been influenced by Foucault's work on discourse, knowledge and power, which we discuss below.

The foci in the anthropology of development on discourse are linked to debates within anthropology concerning its role in reproducing colonial forms of authority and orientalism which came to the fore in the 1970s and 1980s. These critiques questioned the discipline's portrayal of an ahistorical, exotic 'other' which exists in opposition to the Western self. In contrast, within 'postmodern' anthropology all domains were seen as valid subjects for research; institutions and discourses from the anthropologist's own society become relevant areas of study (Marcus and Fischer, 1986: 111–13); To redress the balance of previous orientalism, it was suggested, anthropologists should deconstruct cultural assumptions of the North as well as the South – what Rabinow terms 'anthropologising the West' (1986: 241). Such work can also indicate how power is gained, and reproduced, at local, national and global levels. While there are many potential fieldwork sites for this, 'development' was an obvious candidate. This might involve studying aid agencies, the categories, knowledges and culture of development, or conducting fieldwork among expatriate groups.

To understand what was meant by 'development discourse', we should start with the work of Foucault, arguably the most important thinker of the late twentieth century. In *The Order of Things* (1970), Foucault focused upon 'fields' of knowledge, such as economics or natural history, and the conventions according to which they were classified and represented in particular periods. While they are represented as objective and politically neutral, he shows how areas of knowledge are socially, historically and politically constructed. Discourses of power, while presented as objective and 'natural', actually construct their subjects in particular ways and exercise power over them. Malinowski's 'scientific ethnography', for example, claimed to generate objective and scientific accounts of native 'others', which presented them in a particular light and so justified their subordination. Knowledge is thus inherently political. As Foucault put it: 'the criteria of what constitutes knowledge, what is to be excluded, and who is qualified to know involves acts of power' (1971; cited in Scoones and Thompson, 1993: 12). Discourses thus subsume practices and structures, with very real effects.

From this, areas of developmental knowledge or expertise can be deconstructed as historically and politically specific constructions of reality, which are more to do with the exercise of power in particular historical contexts than presenting 'objective' realities. The notion of discourse 'gives us the possibility of singling out "development" as an encompassing social space and at the same time of separating ourselves from it by perceiving it in a totally new form', as Arturo Escobar (1995: 6) has argued. How such discourses interrelate with other structures, the ways in which they are contested and the interface between developmental and other forms of knowledge are just a few important questions generated by this approach. This is an area where the study of development has a major role to play in wider theoretical debates in anthropology, for development projects provide an opportunity for examining the dynamic interplay of different discourses and forms of knowledge (Worby, 1984).

Escobar became and remains a key figure in the theorising of developmental discourse. In a paper published in 1988, for example, he examines the history of development studies and its production and circulation of certain discourses, seeing these as integral to the exercise of power; what he calls the 'politics of truth' (Escobar, 1988: 431). Development practice, he argues, uses a specific corpus of techniques which organise a type of knowledge and a type of power. The expertise of development specialists transcends the social realities of the 'clients' of development, who are labelled and thus structured in particular ways ('women-headed households', 'small farmers', etc.). Clients are thus controlled by development and can only manoeuvre within the limits set by it. As he put it in *Encountering Development*, 'Development had achieved the status of a certainty in the social imaginary' (Escobar, 1995: 5).

In *The Anti-Politics Machine* (1990) James Ferguson took a similar approach by analysing the Thaba-Tseka project in Lesotho. The resulting text demonstrated the possibilities for project ethnography. Rather than being concerned with whether development is 'good' or 'bad', or how it could be improved, Ferguson argues that we should analyse the relationship between development projects, social control and the reproduction of relations of inequality. This cannot be simply explained by models of dependency; structures do not directly answer the 'needs' of capitalism, but reproduce themselves through a variety of processes and struggles. By analysing the conceptual apparatus of planned development in Lesotho and juxtaposing this with ethnographic material from a project's 'target area', he shows how, while development projects usually fail in their explicit objectives, they have another often unrealised function: that of furthering the state's power (Ferguson, 1990: 13).

The Anti-Politics Machine opens with the deconstruction of a World Bank report on Lesotho. Ferguson shows how its amazing inaccuracies and mistakes are not the result of bad scholarship, but of the need to present the country in a particular way. Lesotho is frequently referred to in the report as 'traditional' and isolated, with aboriginal agriculture and a stagnant economy. In reality this is far from the truth, for the country has long been economically and politically intertwined with South Africa. In addition, the report only considers Lesotho at a national level. The implications are thus, first, that development interventions will transform and modernise the country; and, second, that change is entirely a function of the action or inaction of the government.

Ferguson (1990: 68) argues that discourses are attached to and support particular institutions. Only statements which are useful to the development institutions concerned are therefore included in their reports; radical or pessimistic analyses are banished. The discourse is thus dynamically interrelated with development practice, affecting the actual design and implementation of projects. In its definition of all problems as 'technical' the discourse ignores social conditions, a central reason why the project fails. Crucially too, development is presented as politically neutral. Instrumentally, however, the project unintentionally enables the state to further its power over the mountain areas which it targeted. Rather than this being a hidden aim of developmental practice, and the discourse a form of mystification, Ferguson argues that development planning is a small cog in a larger machine; discourse and practice are articulated in this, but they do not determine it. Plans fail, but while their objectives are not met, they still have instrumental effects, for they are part of a larger machinery of power and control.

Considering development as discourse raises important questions about the nature of developmental knowledge and its interface with other representations of reality. Anthropology can have an important role here, first in demonstrating that there are many other ways of knowing (thus undermining development's hegemonic status), and, second, in showing what happens when different knowledges meet. In another contribution, for example, the relationship between scientific and local knowledge within development practice is explored. As the articles in *An Anthropological Critique of Development* (Hobart, 1993) indicated, claims to knowledge and the attribution of ignorance are central themes in development discourse. The scientific and 'rational' knowledge favoured by development constructs foreign 'experts' as agents and local people as passive and ignorant.

Rather than presenting local knowledge as homogeneous and systematic, these accounts showed that it is diverse and fluid. These multiple epistemologies are produced in particular social, political and economic contexts; instead of being bodies of facts, what is important is how, rather than what, things are known. This was a different approach from much of mainstream development discourse, where knowledge is only mentioned as an abstract noun, and those who know are thus stripped of their agency (Hobart, 1993: 21). It was also tied to a growing critique of the 'farmer first' movement, which, while providing a necessary corrective to modernisation theory's assumption that traditional beliefs and practice are an obstacle to progress, tends to simplify and essentialise local knowledge, or assume that, like scientific knowledge, it can be understood as a 'system' (Gatter, 1993; Scoones and Thompson, 1993, 2009).

Within these accounts people appeared as agents, whose knowledge interacted in a variety of ways with that of development agencies. Richards, for example, showed how, rather than being free-standing, indigenous knowledge can be understood as improvised performance. West African cultivators possess performance skills as well as technical and ecological knowledge, mixing their crops in a certain way, providing food and drumming for their labourers, and so forth. This has been missed by most agricultural research and its ensuing 'scientific' expertise, which carries out agricultural experiments in 'set' conditions, ignoring the vital fact that farmers use their creativity and performance skills in cultivation (P. Richards, 1993).

In other words, people do not passively receive knowledge or directions from the outside, but dynamically interact with it. Another example of this was provided by Burghart (1993), who set out to study local knowledge of health and hygiene in a Hindu cobblers' village in Nepal. Although Burghart assumed that there would be a symmetrical exchange of knowledge (his technical knowledge versus their views on hygiene) and that he could construct an objective model of their knowledge, this was not to be the case. Instead, the cobblers refused to accept his role, constructing him instead as a Hindu lord, who was seen as benevolent when the well-cleaning he had initiated went well, and then as malevolent when the water became bitter.

As this body of work indicates, anthropologists need to examine the ways in which people and the discourses which they produce interact according to their different cultural, economic and historical contexts. Research must be actor-oriented, not only through studying those to 'be developed', but in terms of how individual and group agencies cross-cut, reproduce or resist the power relations of state and international development interventions (see also Grillo and Stirrat's edited volume, *Discourses of Development*, 1997). Through these and similar insights, the anthropology of development opened up and became something infinitely more interesting than simply the study of the 'problems' of development.

It does, however, raise some troubling problems, which we grappled with in the 1996 edition of the book. If development is to be understood as a hegemonic discourse in which Third World peoples are objectified, ordered and controlled, how could anthropological involvement in it be justified? Surely the only ethical response was to vehemently reject it and walk away? While accepting that development was politically highly problematic, in 1996 we argued that non-involvement was not the only possible response. Instead, we showed how there were various important ways in which anthropologists, the methods they use and the insights they have could help subvert and reorient development, contributing to its eventual demise and transformation into post-development discourse. We shall be returning to these debates in the next chapters. In the second part of this chapter, however, we bring the anthropology of development into the twenty-first century, where, as we shall see, the field is flourishing.

The anthropology of development from 2000 onwards: new agendas, old questions

So, what happened next? While it is rarely possible to draw sharp lines of demarcation around research areas, by the twenty-teens we can discern two loosely interlinked foci. The first speaks most directly to 'development' as a discursive field. In this work, the practices, representations and framings of the aid industry and its projects are deconstructed and analysed. Rather than being a largely unknown and homogenised entity which appears in scare quotes, this new ethnography of aid adds detail to development. Its contradictions, conflicts, personalities and unintended consequences (Ferguson, 1990) spring to life in the form of 'Aidland' (Apthorpe, 2011) and Aidnography (Gould, 2004).[10] While welcoming this work and its contribution, we will also suggest that Aidnography has reached its own impasse, and is in danger of maintaining an overly narrow focus which fails to address the more urgent questions thrown up by the fast changing global context.

In the second area, 'development' fades into the background and the research focus falls upon 'little d' development: economic and social transformations, the ways in which the world, its global and local flows, cultures and relationships are changing. Some, if not all of the anthropologists working in this field might baulk at the suggestion that what they are studying is development. But as we shall argue, whether labelled 'Millennium Capitalism' (Comaroff and Comaroff, 2000), 'neoliberalism', 'new economic orders' or good old-fashioned development, the study of our changing and conflicted world, with old and new struggles, injustices, and winners and losers, is increasingly inescapable. Rather than a marginal 'applied' interest for the less scholarly, or anthropology's 'evil twin' (Ferguson, 1997) it has become part of the beating heart of the discipline, core to its moral purpose (Gow, 2002; Edelman and Haugerud, 2005; Edelman, 2013) and inescapable in its newly invigorated ethics. As Olivier de Sardan argues, it (development/Development/'development') is 'neither an ideal nor a catastrophe. It is above all an object of study' (2005: 25).

Rationalities, technologies and the production of success: ethnographies of the anti-politics machine

First, though, back to 'Development'. For all the critiques of Escobar and other deconstructionists for their 'caricature or *reductio ad absurdum*' of development (Olivier de Sardan, 2005: 5), the discursive turn has brought huge insights and opened up important new areas of research, not to say sharp tools for radical critique, revealing how the 'anti-politics' machine operates. In their revisiting of what had originally been powerful feminist concepts for example, Cornwall and colleagues (2007: 2) show how development institutions, the power relations they involve and the myths they create, 'undermine feminist intent', co-opting and depoliticising at every turn. In a similar vein, Cornwall and Eade's (2010) volume on 'fuzzwords and buzzwords' explores how originally radical ideas involving participation, empowerment and partnership have been standardised and emptied of their original purpose by the anti-politics machine. We shall return to this work and its implications in Chapter 5. Meanwhile there have been important ethnographic contributions that further demonstrate workings of the 'anti-politics' machine. Ferguson's (1990) early ethnography of a World Bank livestock project in Lesotho, discussed in the previous section, was a precursor to an influential ethnography published 17 years later: Tania Li's *The Will to Improve* (2007).

In her analysis of colonial, state and donor improvement schemes in Indonesia, Li concurs with Ferguson that such schemes must be understood in Foucauldian terms as forms of governance which erase politics and evade questioning or scrutiny. Unlike explicit forms of discipline, governance operates by educating desires, aspirations and beliefs, which people often don't notice happening (Li, 2007: 5). By rendering what are essentially political problems such as extreme poverty, or the loss of livelihoods to industrial development as 'technical' issues with a range of technical solutions, projects of 'improvement' govern by the back door. Li argues that a key part of the process is 'problematisation', whereby a problem is identified and a solution offered. Another element in contemporary schemes is the role of experts, who: 'focus more on the capacities of the poor than on the practices through which one group impoverishes the other' (Li, 2007: 7). In Indonesia, Li shows how World Bank 'community development' schemes drew from romanticised notions of a previously existing 'natural' community that the project would restore via particular techniques such as the setting up of committees. Meanwhile poverty was diagnosed by the World Bank team as resulting from undemocratic structures and corruption rather than socio-economic relations, the solution being the introduction of neoliberal practices of competition and accountability: 'To govern through community requires that community be rendered technical. It must be investigated, mapped, classified, documented, interpreted ...' (Li, 2007: 234). Seen in this light 'community development' is an anti-politics machine *par excellence*, distracting attention from the true causes of poverty, offering technical solutions to political problems, and governing, not through discipline and authority but the implementation of new administrative structures, edicts and bureaucratic structures, all underlain by implicit neoliberal ideology. Crucially, Li shows how such schemes are not simply accepted by the people for whom they are intended; governance is limited, for: 'men in their relations, their links, their imbrications are not easy to manage' (Li, 2007: 17). This point is vital, for by showing the difference between the practice of governance and the practice of politics in which the schemes are challenged by Realpolitik, 'development' emerges less as an all-powerful machinery, hoovering up dissent and spreading neoliberalism across its (soon to be) dependent territories, and more as a contested domain, around which struggles take place. To this extent Li has added history and context to Escobar's discursive analysis; her ethnography tells a richer, more nuanced story.

David Mosse's *Cultivating Development: An Ethnography of Aid Policy and Practice* (2005) may at first glance seem similar to Li's *The Will to Improve* for both are essentially ethnographic accounts of development projects and policies. But whereas Li's focus is upon the rationale of improvement schemes and what happens when they become entangled with the real world, Mosse describes what happens within the day-to-day implementation of such projects; indeed, projects are as much 'the real world' as the contexts in which they take place. His story has an intriguing and original twist. Rather than seeing projects – the practice of development – as resulting from policy as a rational solution to problems, Mosse argues that the relationship between policy and practice is more complex. Development work has inner logics and rationales in which the ultimate, though hidden objective is not success per se, but the production and appearance of success. For Mosse the ethnographic question is therefore not *whether* but *how* development projects work, not whether a project succeeds but how success is produced (2005: 8). The black box of implementation is finally opened up.

Using his experience of working within a British-funded irrigation project in India, the Indo-British Rainfed Farming Project, Mosse takes us deep into its everyday logics. What emerges is an account of how the appearance of success is socially produced by the actors involved, be these consultants, project managers or distant policy-makers. 'Participation', for example, is turned into a commodity via skilful PR, part of the project brand; by being promoted as a model of successful participatory rural appraisal (PRA), the project becomes a supplier of this (Mosse, 2005: 159). In order to keep up the appearance of success, VIP visits and publicity material become more important to the daily management of the project than its actual outcomes. Or rather, the appearance of success is the actual outcome, a performance for a particular audience, for projects involve particular conceptual and linguistic devices which inspire allegiance and conceal ideological differences (Mosse, 2005: 12). As Mosse writes: 'In selected villages everyday life gives way to project time, space and aesthetics. The village is organised to resemble the project text so as to be pleasingly read by outsiders' (2005: 165). Through his ethnography of the production of project success, or how development is 'cultivated', Mosse is able to make a number of insightful observations about the nature of development policy and practice. First, policy exists to legitimate practice rather than orient it; second, development interventions are driven by the needs of organisations rather than policies; third, development projects work to maintain themselves as coherent policy ideas; fourth, development projects don't fail 'in themselves' but are failed by their networks of support and validation; and, fifth, 'success' and 'failure' are policy-oriented judgements which obscure project effects (Mosse, 2005: 181).

As we saw in Chapter 2, Mosse was somewhat taken by surprise when *Cultivating Development* caused controversy among some of his former project colleagues in India. The book has, however, been a major contribution to the anthropology of development, opening up development policy and its implementation, as well as the 'hidden transcripts' which lie behind public statements and performances as important areas of enquiry. Rather than culture and social relations existing only for those at the receiving end of development, we see how both lie at the heart of development itself. 'Development' certainly involves discursive power, with the attendant framings, ways of seeing and labelling, but it is nothing like the hegemonic, all-powerful force conjured up by Escobar or Sachs and others' analyses ten years earlier.[11] It is instead just like other areas of human endeavour: contingent, contradictory and not always effective.

Aidnographies of Aidland

Mosse is not alone in his interest in the internal machinations of development. *Cultivating Development* is just one example of what has proven to be a fertile field of study: 'aidnography', defined by Raymond Apthorpe as 'exploring the "*représentations collectives*" by which Aidmen and Aidwomen say they order and understand their work' (cited in Mosse, 2011a: 2). Richard Rottenburg's *Far-fetched Facts: A Parable of Development Aid* (2009) is an interesting example. Focusing on the interstitial spaces between models of development and where they are supposed to be implemented, Rottenburg takes a novel approach to the everyday work and outcomes of development by presenting the shifting positions, understandings and decisions of different experts and consultants as fictionalised narratives. By interweaving their accounts of the progress of a sanitation project in an anonymous African country, one of Rottenburg's main aims is to 'give voice' to the actors he argues are missing in social science accounts of development. Rather than these being the 'natives': 'on the contrary it is the voices of technical experts from the North ... that are rarely heard' (2009: xxxvii).

Over recent years this gap has been filled by two collected volumes which delve deep into the practices, moralities and motivations of those who create policy and carry it out. In *Adventures in Aidland: The Anthropology of Professional International Development* edited by David Mosse (2011a) contributors describe the cultures and identities of aid workers, their everyday worlds and the ways in which their knowledge is constructed and travels across the world, packaged as expertise. This knowledge forms the basis of new orthodoxies which – following Craig and Porter (2006) – assert the universal over the particular, technical priorities over political ones, and spread risk downwards to the places where policies are implemented. As Mosse (2011a: 3) argues in his introductory chapter: 'Perhaps never before has so much been made of the power of ideas, right theory or good policy in solving the problem of global poverty.' While the ethnography of policy is not new (see for example Wright and Shore, 1997), Aidnography shows how development professionals frame and negotiate their expertise within the shifting constellations of institutional politics, perform their professional identities and maintain the paradigm of technical change and progress. Ian Harper's chapter on international health workers in Nepal, for example, shows how spatial practices such as the fortification of the US embassy and USAID offices in Kathmandu allows development experts to lead parochial lives within Nepal while the true cosmopolitans are Nepalese migrant health workers in the UK, who are far more experienced in moving between cultural as well as geographical domains (Harper, 2011). In a similar vein, Rajak and Stirrat (2011) write provocatively of the 'parochial cosmopolitanism' of rootless experts, who mourn the changing worlds they move through, even though it is the application of their expertise in the name of development that has led to those changes.

Writing as a seasoned inhabitant of 'Aidland', Rosalind Eyben adds further nuance to these perspectives. Eyben argues that, rather than simply arising from parochialism or nostalgia, changing practices within the history of international aid have necessitated social networking and the creation of community among aid professionals in their overseas postings; the whirl of drinks parties, picnics and receptions that she experienced while working for the Department for International Development (DFID) in Bolivia are not merely to do with 'culture' but are linked to the functioning of the aid industry. Indeed: 'this donor community can be best understood as a local cell of a global "virtual" international aid community' (Eyben, 2011: 141). Instead of gaining in-depth country experience, aid workers are expected to move smoothly between postings, circulating standardised approaches to development (or Mosse's 'travelling rationalities'; 2011a: 3) in which local context is irrelevant. Meanwhile the 'harmonisation' of best practice between donors means that socialising must include the representatives of other international agencies. In other work Eyben has taken a similarly historical approach, showing how the discursive work of development is dependent upon changing government policies as well as negotiations between interest groups and individuals within the organisation. In a piece detailing the production of booklets on 'Women and Development'/'Gender and Development' over the course of 20 years by DFID,[12] for example, she charts the ways in which representations of women changed as the agenda of women and development was 'mainstreamed' by social development advisers such as herself within the organisation, despite considerable resistance from other quarters (Eyben, 2005).

A second volume, *Inside the Everyday Lives of Development Workers* (Fechter and Hindman, 2011), locates the outcomes of development within the work of Aidland, arguing that the culture and everyday practices of those working at the frontline have been neglected by theories which foreground ideology and policy. Like *Adventures in Aidland* the volume takes us deep into the world of development, putting particular emphasis on the moral worlds and ambiguities of those employed within it. If there is a difference between the collections, it is that contributors to *Inside the Everyday Lives of Development Workers* focus more on the personal experiences of development workers and, in particular, upon the moral and spiritual tensions which beset them. In his chapter on the moral ambiguities facing development workers in Ghana, who negotiate between ideologies of good government and transparency and the exigencies of personal connections in West Africa, for instance, Thomas Yarrow (2011) shows how, like any social world studied by anthropologists, Aidland involves specific moral economies. As we shall see, these questions of morality have become increasingly important to the anthropology of development in the twenty-teens.

*   *   *

There can be no question that Aidnography adds to our understanding of development, shining a light on the rationalities/irrationalities, cultures and histories which shape policy and practice. But although Aidnography has been important and useful we must also move beyond it. Not to do so would mean that we are in danger of missing bigger, more fundamental questions. As aid practitioners/anthropologists (the two are often the same) study themselves, the world beyond Aidland or 'Bubble Land' (Apthorpe, in Mosse, 2011a) faces huge challenges: climate change, global recession, resource scarcity, widening inequalities and violent conflict to name the most obvious. It is these areas, and not the study of aid organisations, that, by the second decade of the twenty-first century, urgently require robust anthropological critical inquiry. By challenging taken-for-granted orthodoxies, complicating simplistic understandings and revealing the everyday realities of those who live through the ruptures and transformations of our fast changing world, new insights may be gained. Vital questions remain which anthropologists are supremely well-positioned to answer. These may seem obvious, even simple, but are, in our view, central to an ethically informed and politically engaged anthropology. Put simply: *in times of change who wins and who loses? How does this happen?*

In the remainder of this chapter we suggest that the most important and exciting work within the anthropology of development is taking place beyond debates surrounding discourse and workings of Aidland. Indeed, Aidland as configured in existing Aidnography seems strangely old-fashioned. Today, multinational corporations, billionaire philanthropists, Islamic charities and new state donors such as China and India are increasingly influential players. Meanwhile, the issues which detain anthropologists of development have exploded into the mainstream. Writing in 2005 for example, Edelman and Haugerud call for an anthropology which engages directly with political economy, positing a list of questions which in their view have been under-explored (and which might be subsumed under the label 'development'). These are: (1) the changing and contested boundaries of the concept of the 'market'; (2) the persistence of 'moral economies' despite the rise and rise of pro-corporate free market policies; (3) the invention and trading of new commodities; (4) grassroots groups which resist and provide alternatives to liberalised economies (Edelman and Haugerud, 2005: 21). Though there are other issues to add, including global finance, contestations surrounding the use of natural resources and the role of corporations in reshaping the world, as well as new approaches to feminism, sexuality and development, Edelman and Haugerud's list is remarkably prescient for, by 2014, some of the most exciting anthropological work is located within these areas. Rather than being corralled off as 'the anthropology of development', what has emerged is a newly invigorated anthropology in which engagement with global change, via a muscular analysis of the relationship between economy, ideology and culture, is no longer a specialist area and the distinctions between 'pure' and applied anthropology seem absurd. While the focus of some has been on the development industry, the new anthropology of development has a wider remit, encompassing work on trade and markets, corporations and ethical business, global finance, feminist and other grassroots movements, and environmentalism. As we shall see, the relationship between academic anthropology and activist anthropology (as opposed to an apolitical 'applied' anthropology) is increasingly blurred.

The 'development gift', morality and belief

While global crisis, rapid economic transformation and the emergence of new donors may be the context for a revitalised anthropology of development, a shift in theoretical perspective is another. If Foucauldian analysis contributed to a 'postmodern challenge' in the 1990s, reinvigoration has come from an unlikely quarter: classical anthropology in the shape of Mauss's theory of the gift. While it would be misleading to state that this has reshaped the field analysis of development encounters as a form of exchange rather than merely as oppressive hegemonic control, it has allowed some anthropologists to connect the study of development to core concerns within economic anthropology. Crucially, it has helped to open up an important new area of enquiry: the role of morality and spirituality. Just as anthropologists such as David Graeber (2011), Gillian Tett (2010) and Karen Ho (2009) have used anthropology to radically decentre orthodoxies concerning the rational nature of the market and finance within neoliberal economics (see also Elyachar, 2012; Schwegler, 2012). Anthropologists in the last decade have analysed development not simply as an exercise in global or local power relations but also as moral projects, in which ethics, meaning and belief lie at the centre of the action rather than in a silo marked 'culture'. This shift in emphasis is linked to the increasing dominance of emerging players in Aidland – such as philan-thro-capitalists, multinationals touting corporate social responsibility, celebrities turned humanitarians/activists and religiously based NGOs – for which ethical conduct and the moral virtue of giving are stressed.

Why should a theory of gift exchange first published in 1925 be relevant to the anthropological study of development 90 years later? The answer is that by understanding 'the development gift' as invested with social and spiritual meaning as well as being embedded in power relations (Stirrat and Henkel, 1997), anthropologists' use of Mauss has helped them de-centre conventional understandings of aid, charity and projects of progressive change as resulting simply either from a benevolent 'will to improve' or, indeed, from a Foucauldian 'will to control'. Like all good anthropology, the result is a deepening and complicating of what at first sight appears obvious. To this extent the political project of the new anthropologist of development is to kick the homilies and assumptions aside and reveal what lies behind them.

In *The Gift* Mauss argued that gift exchange within 'archaic' societies, which involved seemingly irrational practices such as potlatches among North American Indians or the Kula ring among Melanesian islanders, was in fact an intrinsic social form and indeed the basis of economic life, creating relationships between groups, and was part of a 'total system' which pervaded economics, religion, law, morality and aesthetics. Centrally, gift exchange created the obligation of reciprocity. It also created inequalities between givers and receivers: to give is to assert power over the group which receives. Mauss further argued that gift exchange would break down with the advent of modernity. In his schema, societies move from the 'archaic' 'system of total services' (Mauss, 1990: 5–6) based on gifts, to a modern, fragmented society in which gift giving is superseded by commercial relations, an argument which more recently has received substantial reworking (see Parry, 1986).

What has this got to do with development? First, aid can be seen as a form of gift exchange, whether between states or between individuals donating to NGOs which purport to bring development or other assistance to Third World others. Second, since we know from Mauss that gift giving is both deeply moral and foundational to the creation of inequality, we can analyse development transactions in these terms. Thus, in their seminal article 'The development gift', Stirrat and Henkel (1997) argue that, despite international development orthodoxies of 'partnership', with their implications of equality and sameness, what development gifts actually do is mark difference and hierarchy. By tracing the biography of the development gift, the authors demonstrate how the 'pure' charitable gift is transformed through its journey from Northern donors to Southern NGOs into 'the currency of systems of patronage' (Stirrat and Henkel, 1997: 74).

The use of gift theory thus shows how aid, charity and humanitarianism are essentially social acts, informed by ideas of morality and ethics but grounded in unequal power relations (see Yeh, 2013; Scherz, 2014). These insights are related to a growing interest in the spiritual dimensions of charity and development work. In her ethnography of Protestant NGOs in Zimbabwe, for example, Erica Bornstein shows how religious ideas intersect with the moral dynamics of development work, arguing that, as Clifford Geertz asserts, religious beliefs are 'models for lived reality' and should be understood as informing all areas of life, including the planning, implementation and reception of development projects, an area which until recently has received remarkably little attention (Bornstein, 2005: 2–3). As she concludes with reference to Zimbabwe:

A primary concern of Weber's thesis on protestant religion and rationality is how the aesthetic ideal of methodical, rational and careful action leads to the generation of wealth in this world and how this rational logic gets unhinged from its religious origins, leading to the rise of instrumental rationality and the supposed disenchantment of the world. Protestant NGOs involved in development projects are particularly suited to address this tension, between enchantment and the material improvement of lives. (2005: 4)

Bornstein's work raises fascinating questions concerning the religious and moral dimensions of development, indicating that there is more involved than simply spreading the techniques of neoliberal governance. More research is required in what promises to be a fertile field, including work on Islamic development agencies informed by Muslim practices of giving (Benthall and Bellion-Jourdan, 2009; Benthall, 2010), NGOs based on Hindu or Buddhist principles (Bornstein, 2012; Watanabe, 2013) and work on other non-traditional donors such as the Chinese. Meanwhile, the moral underpinnings and cosmologies of humanitarianism and philanthropy are the subject of growing anthropological interest (see Fassin, 2010; Bornstein and Redfield, 2010; Redfield, 2012; Hopgood, 2006).

How do the moral and spiritual worlds of those at the receiving end of development gifts and interventions inform the results? More generally, how do systems of meaning and belief intersect with discourses of progress and modernity? In previous eras these questions might have been answered by recourse to the binaries of tradition versus modernity, or indeed scientific versus indigenous knowledge. Recent work has broken down these distinctions, showing not simply how people move between apparently oppositional positions but also how what seem to be 'traditional' or non-modern practices are used to critique forms of development which are longed for but fail to materialise. Daniel Jordan Smith's compelling ethnography of practices and discourses of corruption in Nigeria, for example, shows how Nigerians live simultaneously in two worlds in which, on the one hand, the expectations of reciprocity and patronage remain, while on the other the ideals of modern governance and citizenship are aspired to. As Daniel Smith writes: 'everyday practices of corruption and the narratives of complaint they generate are primary vehicles through which [people] imagine and create relationship between state and society' (2007: 6). James Howard Smith's Kenyan ethnography *Bewitching Development* (2008: xii) focuses directly on 'the development imagination as it unfolds and is materialised in a specific region', and, in particular, its dialogic and shifting relationship with witchcraft beliefs and accusations. Arguing that despite the dismissal of development by deconstructionist anthropologists, the ideals of progress and modernity remain compelling for people in much of the world, James Howard Smith demonstrates how the notion of development is *relational* and, as such, constitutes an important means of comparison with other groups and places. Moreover: 'development becomes a prism for reimagining order and progress when established mechanisms for achieving development – such as state patronage and formal employment – have been thrown asunder' (2008: 4). In Kenya, James Howard Smith shows how development is not oppositional to witchcraft but operates in tandem with it; witchcraft accusations become a repository for all that needs to be rejected in the future, a practice that arises both as a result of localised social relations and the failings of the state (2008: 5). Analysing 'witchcraft' and 'development' as unstable and historically contingent concepts, this nuanced ethnography reminds us that development is first and foremost a system of meanings, constantly reworked by the people who invoke it and occupying an ambiguous position in how history and the future are imagined.

Corporate social responsibility, ethical business and moral markets

As noted above, the current anthropological interest in the ethics and morality of development is partly a response to the appearance of businesses and corporations touting policies of corporate social responsibility (CSR). Today, corporations and billionaire individuals alike seem to feel duty bound to develop their own foundations and projects aimed at bettering the world or somehow alleviating the ills of capitalism. While there is much debate over whether this is just more 'green wash', interesting questions remain as to why businesses now need to claim the moral high ground and what effects this has (see for example, Rajak, 2011; Jenkins, 2005; Burton, 2002; Kapelus, 2002; Welker, 2009; Kirsch, 2010a, 2010b; Gardner, 2012). Here, Stirrat and Henkel's insights on the 'development gift' have proven rich in analytic possibilities. Dinah Rajak, for example, has used gift theory to analyse CSR programmes carried out by the Anglo-American mining company in South Africa. Rather than modern business and markets being amoral, as suggested by Mauss and indeed conventional understandings of neoliberal economics, Rajak argues that CSR brings morality into business practice, allowing mining companies to extend moral authority over the places where extraction takes place via moral discourses that stress partnership, responsibility and so on. Not only do global codes of ethics act as a form of governmentality (Dolan, 2007), they also naturalise neoliberal tenets such as entrepreneurship and the role of the market while eliding questions of power and ecology. Instead of being a 'moral bolt-on' to offset the harsh realities of neoliberal capitalism, CSR is therefore intrinsic to its workings. Indeed, the gift of CSR allows transnational corporations to remain moral while at the same time propounding the extension of global markets and business in the name of their moral code. While CSR discourses of 'partnership', 'empowerment' and 'participation' allow the company to avoid charges of patronage, on the ground the politics of the gift remain, sometimes forging alliances and reducing conflict with the people surrounding the mine, but always with the power relations of giver and receiver intact (Rajak, 2011).

Other anthropologists probe the ethical claims of corporations but like Rajak resist simplistic denouncements that CSR 'doesn't work' to reveal the hidden rationalities and contradictions involved (see, for example,

Shever, 2010; Rogers, 2012; Welker, 2014). Jamie Cross, for example, has described how ethical accounting regimes in the De Beer's global diamond industry allow the company to follow a 'corporate ethic of detachment' at sites of manufacturing such as the South Indian industrial unit where he did fieldwork (Cross, 2011). In a similar vein, Gardner argues that for the multinational energy company Chevron, discourses of 'sustainability' and 'helping people to help themselves' allows the corporation to claim they follow the ethics of partnership in their CSR programmes in Bangladesh, while simultaneously following the ethics of detachment (Gardner, 2012). Welker's work on the Newmont mining development in Batu Hijau, Indonesia, raises important questions regarding not only the instrumentality of CSR for corporations seeking moral legitimacy in carrying out environmentally destructive mining projects but also its relationship with the exigencies of security. In Batu Hijau, Welker shows how local elites, already organised into NGOs and civil society organisations, were co-opted by Newmont. Since these groups gained control of the community development benefits offered by Newmont it was in their interest to repress violently anti-mining protests by environmental activists. While Newmont is a signatory to the Global Compact and mine guards have been given human rights training, Welker (2009: 146) argues that the CSR programme has 'produced fresh zones of struggle and new forces of violence' in the area. The violence against the protesters was carried out not by the mine guards but by the local elite, who were defending their interests in the benefits and 'development' offered by Newmont's CSR programme. Meanwhile, by eliding themselves with the environmental movement, CSR managers were able to undermine the claims of the protesters, using the discourse of criticism to counter-attack the protesters. We shall return to anthropological critiques of mining projects shortly.[13]

An important element of the 'moral turn' in business practice is that to be seen to do good has become an important sales tool for consumption and has become invested with moral meaning in wealthy societies in the North. Rather than action against the injustices of global capitalism being a collective effort it has become individualised, turned into a consumer decision about what clothes, vegetables or coffee beans to buy, or indeed what companies to buy shares in. It is not simply that corporations should be 'doing development' in addition to their business activities. Rather, the terms of trade and conditions of employment of producers in the global South have become an ethical domain. Development, whether in the form of projects which aim for fairer terms of trade or ethical codes imposed by the Northern businesses contracting manufactures in the South, has thus become enmeshed with the market, which is invested with moral power, and seen as the remedy to global ills. This new emphasis on 'the market' as a cure-all for global poverty runs parallel with the myth that it is 'free' and/ or governed by rational forces which, if allowed sufficient incentives and investments, will naturally bring progressive change and development. In recent years an important body of anthropological work has challenged these paradigms, showing how neoliberal finance and markets are far from rational but deeply imbricated in social relations. In particular, and as we saw in Chapter 2, Karen Ho's ethnography of the stock exchange (2009), Gillian Tett's account of global financial crisis (2010) and David Graeber's work on debt (2011) demonstrate the power of anthropology in critiquing received thinking about economics as well as developing a public anthropology that contributes in important ways to global debates.

In an influential volume titled *The Hidden Hands of the Market* (De Neve et al., 2008b) the authors delve into the moral and political relationship between producers and consumers, showing the contradictory pulls of what is less a single social movement and more a range of approaches in which the market is conceptualised in various ways, some overlapping and some oppositional. The Fair Trade movement, for example, involves a critique of the neoliberal market which, it is argued, causes deepening global inequalities if left to its own devices. Meanwhile the notion of ethical business or CSR is premised upon an assumption that inclusion in global trade and markets is inherently good.

Central to all approaches is a conversation about the ethics and morality of economic conduct (De Neve et al., 2008a: 2). In their introduction the authors identify four themes:

(1)   Emphasis upon the role of social relations rather than the impersonality of alienated market relations. For example, Northern consumers wish to know where goods have come from or that the production process has not been exploitative via images of producers on packaging or 'rug-marks'. This may in turn lead to new forms of fetishism, for example in the idea of authentic culture in promotional literature for eco-tourism or family coffee producers in Central America. Meanwhile consumer disgust at 'sweat shops' or child labour have led to codes of conduct for businesses like Primark which operate in places like India or Bangladesh.

(2) Ideas about boundedness and separation which critique the open
     market and aim to bring producers and consumers together, for
     example in the idea of locally produced food.
(3) Arguments concerning fair prices.
(4) Moves towards regulation.

Anthropologists working on the latter show how regulatory frameworks
which in the perspective of Northern consumers inserts ethics and
culturally specific moralities into business conduct can operate as modes
of governance in the contexts in which they are carried out and, as such,
may be experienced in less favourable ways by those subjected to them.
Catherine Dolan's work on Fair Trade flower production in Kenya, for
example, indicates that while Northern consumers quite literally buy into
the ideals of partnership and equality, Kenyan producers understand the
relationship with buyers of the flowers as involving patronage and charity
(Dolan, 2008). Meanwhile Geert De Neve's research on textile workers in
Tamil Nadu demonstrates how international codes of conduct generate
new regimes of power and inequality between different actors in the
supply chain (De Neve, 2009).

There is a final twist to our tale of the market. Quite simply, this is the
notion that rather than the market being the panacea (or cause) of the
poor's ills, the poor *are* the market. By selling products and services aimed
specifically at people who live on less than $2 a day, the argument goes,
venture capitalists not only reach a huge, untapped market and thus access
the 'fortune at the bottom of the pyramid', but simultaneously help the
world's poor by offering a slice of the pie (Prahalad, 2006). The solution is a
new vision of inclusive capitalism that 'works for everyone', fighting poverty
with profitability via 'creative capitalism' as Bill Gates would have it.[14] The
approach is linked to the global success story (though not necessarily the
actual success) of micro-credit as a means to eradicate poverty. Pioneered
in Bangladesh by Muhammad Yunus of the Grameen Bank, micro-credit
aims to create entrepreneurial individuals via small-scale loans and savings
groups, usually involving rural women. Once again, the onus is upon the
rationalising force of finance and markets which, left to do their good
work, will create development from below, an endeavour that anthropologists have treated with considerable suspicion, showing how micro-credit
can push women into debt and dispossession (see, for example, Karim,
2011). Recent years have seen a proliferation of financial 'technologies'
with similar aims, a 'financial inclusion assemblage' with a particular set

of rationalities, subjects and practices, such as micro-finance lent online to community savings groups on global sites like Kiva (Schwittay, 2011).[15] With their neoliberal stress on self-help and entrepreneurship in the context of an absent or 'rolled back' state, these approaches are similar to the small-scale entrepreneurial projects that Julia Elyachar observed in Cairo, described by her as 'anti-development development' (2002; see also 2005). We discuss this work in more detail in the Chapter 5.

While micro-credit institutions offer financial products, others sell goods aimed specifically at the poor, often with the assumption that services which, in other eras, the state would have at least aspired to provide – access to electricity, health services, clean drinking water – are absent. Smoke-free stoves, straws which purify water[16] (Redfield, 2012), health-giving yoghurts, anti-bacterial soap and low-energy laptops feature large. As Jamie Cross and Alice Street note, the language of philanthropy has almost entirely disappeared from CSR initiatives, replaced by a language of ethical business practice in which the pursuit of profit is in itself a developmental solution (Cross and Street, 2009: 5). Here, it is *consumer goods* not the state or development agencies which create health and wellbeing; development problems have been converted into consumer need. In India, for example, an everyday commodity such as Lifebuoy soap has been presented by Unilever as a social good, capable of offering a solution to public health problems as well as delivering profits for shareholders. As Cross and Street conclude: 'The corporate search for value at the "bottom of the pyramid" is transforming the landscapes in which many anthropologists work. What will the pyramid look like when anthropologists look up?' (2009: 9)

Catherine Dolan is one of those anthropologists who have been 'looking up' from the bottom of the pyramid. In her and Mary Johnstone-Louis's research on Avon's marketing techniques in South Africa the authors describe how Avon has been at the forefront of 'uplifting' women from poor communities in order to create a cadre of entrepreneurial and self-reliant 'Avon Ladies'. Selling becomes a moral mission, and empowerment a calculative process (Dolan and Johnstone-Louis, 2011: 27). As they conclude, this figure is:

> purposely engineered, whetted through corporate practices that aim to graft market rationality onto all spheres of life. Indeed, becoming one of Avon's 'empowered entrepreneurs' requires not only a shift in women's work lives, but a redrafting of their physical, spiritual and social worlds

for instrumental ends' (p.30). Via long distance corporate idioms of self-invention and self-discipline, particular subjectivities are produced. Empowerment is 'outsourced' to the bottom of the pyramid, but only for those individuals who have sufficient resources and self-discipline to reinvent themselves. (2011: 31)

Corporations, accumulation by dispossession and activist anthropology

While some multinational corporations position themselves as agents of development, bringing lifesaving goods to the world's poor in their role as humanitarian capitalists (Redfield, 2012), anthropological critiques open up important questions concerning the ways in which particular subjectivities, representations and forms of governance are produced as hidden transcripts in the new 'creative capitalism' which are unlikely to make it into the marketing literature. In other contexts, corporations have more trouble in positioning themselves in such a positive light (though their CSR programmes clearly attempt to do this). This is particularly the case when the corporations require resources such as land, water, gas or minerals for their operations. As political geographer David Harvey has described, the last 20 years have seen their increasing encroachment into new territories in a process he calls 'accumulation by dispossession', in which venture capitalism is pushed ever further from centres of accumulation in its search for profitability (Harvey, 2005).[17] Increasingly, bitter and often violent struggles over resources are the hallmark of economic development in the areas where anthropologists work. Pitted against multinationals and states alike, local groups often become linked to national or global networks in which uneasy alliances between varying interests occur, as in the case of fierce resistance against the planned Phulbari mine to be operated by Asia Energy in Bangladesh, in which local struggles over land are subsumed by nationalist anti-imperialist sentiment at the national level (Nuremowla, 2011). Often framed in terms of 'indigenous' rights over natural resources, analysis of these struggles and their outcomes is an urgent task for the anthropology of development. The old but enduringly vital questions reappear: Who wins? Who loses?

Anthropologists are supremely well placed to answer these questions, offering detailed observations of the effects of neoliberal policies and global capitalism in contexts where industrialisation and urban migration have taken hold. In South Asia, Harvey's 'accumulation by dispossession' has proven to be a compelling formulation for those researching or protesting against predatory global capitalism, galvanising work on agrarian transformation, industrialisation, the formation of labouring classes and the dialectic between formal and informal sectors of production and groups of labourers, often adding important provisos which point to the importance of cultural context, local and national politics and history (for example Münster and Münster, 2012; Walker, 2008; Guérin, 2013; Sanchez, 2012).[18]

Indeed, answering Burawoy's (2000) call for a 'grounding' of neoliberalism, the emerging anthropology of economic transformation aims, in the best anthropological tradition, to connect micro with macro processes, to historicise and complicate analytic terms which imply there has been an epoch-changing rupture (see Münster and Strümpell, 2014; Strümpell, 2014a; Carswell and De Neve, 2014; Corbridge and Shah, 2013). Christian Strümpell, for example, argues that the ways in which different groups of workers react to large-scale job losses in a steel plant in Rourkela, Odisha, are rooted in the politics of the postcolonial regional state since the 1950s (Strümpell, 2014b). In South India Carswell and De Neve and Carswell (2014) show how, in Tamil Nadu, neoliberal transformations, which include an explosion of garment factories, have not led to unilinear changes among different groups of Dalits, but depend upon local differences in inter-caste relations, and their relationship to wider institutions and policies.

While anthropologists of South Asia have tended to focus upon the dialectic between industrialisation, global capitalism and new (or old) forms of inequality, as well as changes in relations between groups, others have focused upon the spread of corporate interests over particular regions. Taking issue with James Scott's (1998) assertion that global capitalism creates homogeneity in the same manner as large-scale state projects of improvement, for example, James Ferguson (2005: 377) argues that it functions quite differently, creating new forms of order and disorder. In Africa, most capital investment has taken the form of resource extraction, in which enclaves are created around mining sites. Often heavily guarded by security forces, the enclaves exist as islands of modernity and control geared around extracting what is needed from 'useable' Africa. Meanwhile 'unuseable Africa' continues to be held together (or not) by NGOs that cover the cracks left by failing or absent states (Ferguson, 2005: 380). The enclaves of order and security created by mining companies, and what this means for the workforce and surrounding populations, are emerging sites of anthropological research (see for example, Yessanova, 2012). If mining enclaves need security and order, they also need moral justification. As noted earlier, most multinational corporations working in the global South now engage in the discourses of CSR in their dealings with local people. The extent to which these practices are primarily to do with boosting their international reputation as well as gaining local compliance in potentially violent contexts is an important question. Zalik's research on Shell Nigeria, for example, shows how, while CSR programmes aim at gaining community consent, the ideals propagated by CSR are at variance with the company's need to maximise profit (Zalik, 2004).

While the anthropology of resource extraction has its roots in earlier work on mining in the Andes conducted by June Nash (1979) and Michael Taussig (1980), in the contemporary period it tends to be focused less upon the meanings and rituals of the miners and more upon the effects of large-scale resource extraction by states and multinationals. A recent volume edited by Behrends et al., *Crude Domination: An Anthropology of Oil (Dislocations)* (2011), brings together anthropologists researching the political economy, meanings and conflict surrounding oil extraction in different parts of the world. Anthropological engagement is often passionate; with the sides sharply delineated and the stakes high – dispossession and environmental devastation often face peoples whose land is to be taken for mining – researchers often become directly involved as activists against multinationals or advocates for local rights: Suzanne Sawyer in Ecuador, Stuart Kirsch in Papua New Guinea and Felix Padel in Orissa (Padel and Das, 2011), India are leading examples. In her ethnography of indigenous struggles over the extraction of crude oil in Amazonian Ecuador, Suzanne Sawyer (2004) charts the development of the indigenous movement, its opposition to the multinationals and neoliberal state, and the discourses deployed against it.

Rather than simply involving struggles over access to resources in which meanings and phenomenology are shared, anthropological analysis reveals how different forms of knowledge often underlie the form that contestations take. In Papua New Guinea, Stuart Kirsch (2006) has charted indigenous responses to the wide-scale destruction brought about by the Ok Tedi Mine, showing how, among the Yonggom people, holistic analyses of the relationship between people and the environment at variance with 'scientific knowledge' underscore their analysis of their situation and political responses to it. Gardner's (2012) research in villages surrounding Chevron's gas extraction site at Bibiyana in Bangladesh throws up similar issues. While Chevron celebrates developmental success and the safety of their high-tech installation via quantitative data, which they posit as 'hard' evidence, local peoples' stories of environmental damage and the loss of livelihoods are dismissed as 'hearsay' by the company; knowledge, and the hearing of particular narratives versus the silencing of others remain central to Chevron's PR operations (Gardner, 2012).

As we argued in Chapter 2, and as the work of anthropologists such as Kirsch, Sawyer and Padel demonstrates, in these contexts the attempt to maintain a rigid distinction between research and activism or 'scientific'/'objective'/'pure' anthropology versus its evil twin, 'applied' anthropology, is ethically questionable.[19] Stuart Kirsch, in particular, has argued that, when faced with dispossession or violence, an ethical anthropology must take sides: 'There is no neutral act of observation. I could either report on the truth or spy for their enemies' (2002: 68). We agree. As the world becomes ever more divided between the 'haves' and 'have-nots', as economic development leads to rupture and dispossession for those with the least power, and the corporations and entrepreneurs which benefit the most from these changes simultaneously parrot discourses of sustainability, 'creative capitalism' and their contribution to the greater good via programmes of CSR, 'partnership' and 'community engagement', anthropologists of development need to use their skills of critique and ethnographic description to the full.

* * *

In the next chapters we return to our original argument, outlined in the 1996 edition of the book. In this, we suggested, first, that there were three central areas which anthropologists examined when assessing D/ development. These involved questions of access, effects and control. Second, rather than there being an absolute distinction between 'practical' or applied anthropology and academic study, each informed the other. It was possible to see how insights gained from the areas of study outlined above had influenced – sometimes directly – changes in policy and practice. Equally, anthropology had been enriched by the study of development.

In what follows we shall return to these perspectives. In Chapter 4, which is presented in its original 1996 version, we show how the questions of access, effects and control raised by anthropologists working within development led to important insights, which fed back into new understandings of project outcomes. While the case studies are drawn from earlier periods, they remain relevant today. In Chapter 5, we revisit some of the ideas that, in 1996, seemed to be subverting the discourse of conventional development, offering new ways of seeing and doing. As we shall see, while some of these new ideas may have been co-opted and even 'turned bad' in the intervening years, there is still considerable cause for optimism.

4

Anthropologists in Development: Access, Effects and Control

As the previous chapter suggests, one of the most important functions of the anthropology of development is its ability to deconstruct the assumptions and power relations of development. While these debates have mostly been carried out within academic domains, other anthropologists have been working hard with and within developmental institutions to alter policy. Such anthropologists may perform a variety of roles: they may be employed as independent consultants, or as salaried staff; others may be involved with pressure groups which lobby agencies or produce alternative visions of change. Anthropological perspectives and methods which help subvert and transform the dominant discourses of development may also be used by a range of non-specialists.

As we described in Chapter 2, such work is not easy. Since donors and development agencies work within a particular discourse, anthropological insights may easily become distorted and 'hardened' into policies which are then applied unilaterally. Once again, the world is packaged and controlled in a particular way.[1] Anthropologists may also face dire contradictions, for their premises are in many ways inherently different from those of developers. While anthropologists are trained to be cultural relativists, development agencies are usually committed to universal principles of progress. This often involves ethnocentric assumptions about what constitutes desirable social change. Strategies of 'social development' and 'women in development', for example, all involve changing society in ways which may not be 'culturally appropriate'.

In this chapter, we outline the main ways in which anthropological insights can be applied to planned change and policy in order to change the dominant discourse from within. Rather than being wholly monolithic, static and encompassing, we suggest that development work actually comprises a variety of countervailing perspectives and practices, as well as a multiplicity of voices. Developmental decision-making and policy are therefore less simple or homogeneous than one might assume. Anthropologists, along with others, can help to unpick oppressive representations and practices, put different questions on the agenda and form new, alternative discourses. As we shall see later in the book, when we revisit these questions with more recent case studies, these approaches and issues remain as pertinent in 2014 as 20 years ago.

Most of the insights which anthropologists provide are rooted first and foremost in common sense. We are not claiming that they have 'exclusive' expertise which others cannot gain access to. One possibility which we will be exploring later in this book is that local development workers might collect their own ethnography and develop their own anthropological intuitions, a practice which, by 2014, has been taken on board by corporations as well as development agencies, as elaborated in the Conclusion. What we do suggest, however, is that the anthropological eye, trained as it is to focus on particular issues, is invaluable in the planning, execution and assessment of positive, non-oppressive developmental interventions. This is not so much because anthropologists have access to a body of objective 'facts' about any given society, but more that they know what questions to ask and how to ask them. While, in retrospect at least, such questions may appear to be obvious, time and time again, as the failure of so many development interventions testifies, they are forgotten. The three boxes included in this chapter set out examples of the types of questions that anthropologists can usefully ask in the context of issues of access, effects and control.

Below are some of the main issues addressed by applied anthropologists. As we shall see, these are deeply informed by the findings of research-focused anthropology. Again, knowledge for understanding and knowledge for action are inseparable. While these questions are often first raised by anthropologists, we suggest that, ideally at least, development anthropologists should not be in the business of predicting what is 'best' for the poor. In contrast, anthropologists working in development can help facilitate ways for the 'victims' or 'recipients' (depending on one's perspective) to have a voice in the development process, so that ultimately it is they who dictate their interests and the most appropriate form of developmental interventions. The rest of the chapter will be organised around the following themes: (1) access, (2) effects and (3) control.

Access

As anthropological research indicates, economic growth can exacerbate rather than eradicate poverty and exploitation. Colonialism and neo-colonialism have meant that the rewards of capitalist growth are spread unevenly between different parts of the world. This means that policies which promote economic growth, or are based on the notion of 'trickle down', are unlikely to benefit everyone equally, for by definition capitalism promotes accumulation for some at the expense of others. This inequality exists at international and national levels, both of which anthropologists may wish to analyse and comment upon. Access may depend on inequality within communities, between local groups and the state, or at an international level. It should be noted, however, that although some anthropologists have attempted to analyse the relationship between world capitalism and global exploitation (for example, Wolf, 1983; Worsley, 1984), the majority are more accustomed to investigating social relations at the micro level.

Although unequal distribution may appear to be an obvious and crucial issue, planners often forget that in the communities where they are working people's access to resources and decision-making power is rarely equal. This may be due to political naivety, but is also because those who plan from the outside tend to assume that 'the poor' are all the same and thus have the same interests. As all anthropologists are aware, however, most communities are highly heterogeneous. There are also many different forms of inequality: those depending upon constructions of race, gender, class and age are just some of the most basic. Each of these, in turn, is structured and experienced according to the particular cultural, economic and political context. We cannot therefore declare that particular groups are always more disadvantaged than others and must thus be the 'targets' of aid. 'Women-headed households', for example, are indeed often disadvantaged. But they are also not all the same, even within the same cultural context, let alone in different societies (Lewis, 1993; Chant, 2007).

Inequality, and differential access to and control over resources, also exists at many levels within communities. This may involve inequality between different households, whether structured through caste, ethnicity, social status or economic class. All of these factors may also cross-cut, or coincide with each other. Inequality may exist between different kinship groups, thus transcending the boundaries of individual households, or it may exist within households, whether this is in terms of gender, age or particular kinship relations. Combined with this, the exercise of power involves various types of relationship, interaction and social action. If power is defined, after Weber, as the ability to influence events, then clearly it may come through a variety of sources. It may be legitimate ('authority') or unofficial (the ability to influence events informally, perhaps through personal relationships, covert strategising and so on).

In considering who gets what, we must therefore be aware of several key issues. First, while inequality exists in all societies, it is structured in particular ways according to its cultural and historical context. Second, power over resources and decision-making is not always explicit. Even while officially there are equal rights for all citizens, in reality this may be far from the case. It is thus hardly surprising that development interventions so often benefit only particular groups, or end up disadvantaging those it was assumed they would help. To illustrate this, let us consider some case studies which illustrate various levels and forms of inequality, and the ways in which this affects people's access to the 'benefits' of developmental resources.

Case 1: Albania: differential access to rural resources in the post-communist era[2]

We begin with a short case study of Albania, the poorest country in Europe in 1996,[3] in which a strictly isolationist and totalitarian communist regime did its best to eliminate economic inequalities in the countryside in the 40 years before 1990 through the imposition of a system of collective farming.

The Stalinist government of Enver Hoxha was repressive and inefficient, but it did meet people's basic material needs and included a comprehensive welfare system which provided reasonable health care and education facilities for the majority of the population. In agriculture, despite low levels of production and a serious disregard for long-term environmental issues, farming inputs such as tractor ploughing services and fertilisers were available and agronomists were on hand to advise the cooperatives.

In 1990, after the upheavals in the rest of Eastern Europe, the government was finally brought down through largely peaceful protest. The political system collapsed, ushering in a new era of social democracy and tentative capitalist development. During the downfall of the government there was a spontaneous and violent uprising by the people, not against the communists themselves but against all the physical trappings of the old regime. Village schools, health centres and other elements of infrastructure were destroyed by angry villagers.

A long period of structural adjustment began, managed by the World Bank and including a privatisation drive, a land reform process and the opening of the country for the first time to foreign investment. But during this period of transition, which, as in most of the former communist countries of Eastern Europe remains in its infancy, most of the services of the former state were in rapid decline or collapsed completely. By 1996 the country was dependent on food aid. The social safety net, which had included a system of old age pensions, sickness benefits and food subsidies, barely exists. Completely unprepared for these new realities, most farmers were thrown back on to their own resources and many retreated into subsistence agriculture. Many villagers returned to pre-communist traditional systems of village government through elders. Local mosques and churches, which had been closed or destroyed under communism, became the community focus for survival and welfare.

A small number of rural people, however, benefited from the collapse of communism by holding on to important cooperative assets at the moment of their dissolution. In one village, the goatherd was able to sell most of the community's cooperative's herd for private gain. In another, a farmer ended up with a tractor which he was able to rent out in a private ploughing service, making enough profit to buy another tractor a year later. Almost overnight, new layers of rural inequality were created; the survival strategies of different households now depending on their level of access to a range of material, social and cultural resources.

Case 2: Mali Sud Rural Development Project: inequality between communities[4]

The Mali Sud Project was launched in 1977 to develop the southern region of Mali – a landlocked country in the Western Sahel. It was extended for a further five years in 1983, funded largely by foreign aid: US$61 million out of a total of US$84 million. The project's objective was to increase agricultural potential in the area by boosting the output of key crops such as maize and sorghum, to promote village development associations and to improve standards of living within rural areas through basic health services and water supplies. The project area spanned some 3500 villages, covering a range of ecological conditions, from arid (having only around 400 mm of rain a year) to the relatively fertile (further south, some areas enjoy 1400 mm of rain annually).

In the first eight years the project did indeed increase the output of many of these crops. Output of the staple foods sorghum and millet increased by 10 per cent, and the area given over to maize saw a 60 per cent increase. But it was only some villages which benefited. In areas where there was inadequate rainfall, maize was ecologically inappropriate. Following the encouragement of the project, however, people had planted maize extensively. In many cases they lost the whole crop.

The main problem with the Mali Sud Project was that it did not help the poorest, many of whom were vulnerable to famine.[5] The project only offered credit and technical advice to farmers who wanted to develop new land and buy new seeds, fertiliser and technology. These were distributed through officially recognised village committees, which not all villages had. Indeed, the committees tended only to exist in wealthier villages, where there was more motivation and organisational skills. Those villages which received help have clearly enjoyed a rise in their standard of living, yet those living in the poorer villages, without a committee, received nothing.

By 1985, some of the poorest villages in the arid areas of Mali which were excluded from the project were on the brink of famine. They desperately needed seeds, especially high-yield varieties to increase their food output, yet were not eligible for help from the project. This was due to two reasons. First, they were not part of a village committee and, second, they had insufficient credit to qualify for a loan with which to buy seeds from the project: all farmers given credit needed at least some capital as a guarantee before being funded. Those most in need of assistance were therefore excluded.

The Mali Sud Project excluded those living in the poorest communities because of pre-determined project criteria. But, as Madeley points out, projects in other parts of the world have demonstrated that the very poor can successfully be given loans without providing material guarantees.

The next case study demonstrates that particular groups can be excluded from project benefits, not because of pre-existing criteria but because insufficient attention has been given to the dynamics of resource allocation in the settlements targeted for 'development'. Rather than simply excluding the most vulnerable groups, this seems to have made their circumstances even more difficult.

Case 3: Land rights in Calcutta: inequality between households[6]

A study of the effects of physically upgrading '*bustees*' (slums) in Calcutta demonstrates how the original, and poorest, inhabitants have tended to be disadvantaged by rather than benefiting from the improvements

(M. Foster, 1989). Slum improvement, which superficially is a physical rather than a social or political process (the provision of sanitation, paved roads, the construction of new houses and so on), thus has variable effects on different groups according to where they are placed on existing socioeconomic hierarchies within the same urban community. Without taking these differences into consideration in the planning stage, and by treating all slum-dwellers as if they have equal access to their homes, Foster argues that such projects have damaging effects on the most vulnerable. As they lead to an unforeseen rise in rents, many of the poorest *bustee* inhabitants are ultimately forced to move to increasingly marginal areas of the city. The upgrading of legal *bustees* has thus been accompanied by a growth in illegal squatter settlements, which are untouched by slum-improvement programmes.

The Indian government had been involved in slum upgrading since the 1970s. In Calcutta, a fund of US$80 million was made available in 1971 to improve environmental and health conditions in the city and, in 1971–81, 1.7 million slum-dwellers were affected by the programme. Important differences in their relative access to property and economic status were largely ignored, however, as were the needs of the poorest of Calcutta's poor, the pavement dwellers. Foster's research into different *bustees* shows a wide range of settlement histories and different types of tenancy among inhabitants. While the earliest settlers often built their own homes, many houses are now owned by landlords who can illegally raise rents through informal *salaam* (gratuity) and key payments, even though officially rents are controlled. Tenants who have moved in more recently tend to pay higher rents; many of these already have jobs in the centre of the city, and some have commissioned space to be reserved for them as their households in other areas expand.

Bustees are thus being 'gentrified' as these richer dwellers move in. Meanwhile landlords are illegally extracting higher rents in a variety of hidden ways. Slum upgrading adds momentum to this process, attracting wealthier inhabitants and enabling landlords to charge more and more. The poorest households, and especially those headed by women, who are particularly vulnerable to landlords' coercive techniques, are thus being forced out.

Foster argues that the key to identifying the beneficiaries of urban environmental upgrading lies in understanding existing patterns of land control. By failing to consider these factors and treating slum-dwellers as all the same, it seems that once again development aids the richest people while disadvantaging the poorest. These effects could only have been avoided by understanding the complex nature of tenancy and property ownership in Calcutta *bustees* at the planning stage, rather than assuming that *bustees* are homogeneous communities, with shared interests.

The negative side effects of slum improvement cannot of course be entirely blamed on *bustee* upgrading. Given the pressure on urban land, such processes are also likely to occur without physical improvements. Avoiding such negative effects is also difficult, for clearly the legal changes necessary for this are beyond the power of urban development authorities or aid agencies. More recent projects funded by foreign donors have not been permitted by local government to work with the poorest pavement dwellers, because they are regarded as illegal squatters. Here, then, constraints imposed by the recipient government have prevented aid from being as 'poverty-focused' as the donors might have wished.

As we know, unequal access occurs within households, as well as between them. In the next case, we shall see how the construction of gender relations in Bangladesh means that even if projects are specifically aimed at women, they do not necessarily benefit from them.

Case 4: Women's credit groups in Bangladesh: inequality within households[7]

In 1975 the Bangladeshi government introduced a programme of rural women's cooperatives in 19 selected administrative districts controlled by the Integrated Rural Development Programme. These women's cooperatives were village-based and structured on the model of pre-existing men's peasant committees. Each cooperative was run by a management committee, elected by members. These represented the cooperative at fortnightly training sessions in health, nutrition, family planning, literacy, vegetable gardening, livestock and poultry rearing, and food processing, sharing their knowledge with other members back in their village. Their primary focus was, however, the granting of small loans, which, in conjunction with the training, was supposed to increase members' income-earning capacity.

In a village studied by Rozario (1992) these loans seemed to be the main reason why women joined the cooperatives. At an interest rate of 12.5 per cent, a woman could apply for 500 taka[8] if she had at least 50 taka worth of shares. Since the interest rates charged by private moneylenders are extortionate in Bangladesh (sometimes running at 100 per cent), and banks are unlikely to give credit to small landowners and the landless, obtaining these loans was obviously highly desirable.

Rozario's research indicated that loans intended to be used by women for their own income generation were either going towards joint household expenses or being co-opted by men. Loans taken out by the poorest women were often spent on basic household items, such as food, clothing and medicine. These women, however, were the ones most likely to invest their loans in growing vegetables or poultry raising. In contrast, wealthier women told Rozario that they did not know how their husbands spent the loans, which they had passed directly to them. They simply signed the forms to collect the loan. Since so many loans were not repaid, with women claiming that they could not control their husbands' decisions or ability to repay, eventually husbands' signatures were required before a loan was made. Men were thus officially given greater control over women's credit.

More recent evidence from elsewhere in Bangladesh suggests similar processes remain common in credit programmes which give loans to women (for a summary see Karim, 2011). Because women and men do not have equal access to resources within households, time and time again loans which are given to women are passed by the recipients to their husbands. Combined with this, because it is women's responsibility to feed and clothe their families, money earmarked for income generation is spent on a household's reproductive needs. Class is clearly an important factor too. Women from richer households, who are more strictly secluded, seem to have the least control over the credit. This may be because ideologies of purdah (female seclusion) prevent such women from entering markets and other public and male domains. The buying and selling of vegetables or poultry may therefore be seen as 'unrespectable' for them, while for poorer women social prestige is not something they can afford. All women, however, shoulder the burden of repayment if and when their husbands default.

By disregarding the ways in which resources are allocated within Bangladeshi households, the cultural construction of women's work and their access to markets, credit programmes in Bangladesh may be controlled by men, even if they are originally intended for women. A key factor here might be that it is cash, rather than other resources, which is loaned. Cash is traditionally associated with male domains, whereas other commodities (poultry, grain, household goods) are traditionally within the female domain. If project planners had located gender relations and inequality within the specific cultural context of Bangladesh, the results reported by Rozario might therefore have been avoided.

Access: Key Questions

What are the most important resources within society?
How is access to resources organised?
Are key resources equally shared, or do some groups have more control than others?
Are there obvious economic differences within communities?

Do some groups have more decision-making power than others?
Are some groups denied a voice?
Are some people incited to speak?
Is access to resources equal within households?
Do some groups have particular interests/needs?

Are there project criteria which constrain some people's access?
Is a certain level of capital necessary ?
Does the project only apply to preconceived categories, e.g. landowners, male farmers or household heads?

Are these factors adequately considered in the development plan/policy?

To summarise, anthropological study of development helps generate a range of questions which focus on people's access to resources provided by planned change. Conventionally in development practice such questions are posed by 'expert' consultants, but this need not necessarily be the case: local participants, activists, non-governmental workers and so on may all contribute. Most important is that the answers are fed back effectively into planning and policy.

Gathering such information is not unproblematic, of course; whether or not the objective 'truth' of socio-political relations can ever be reached is a moot point, not only because outsiders tend to find it extremely difficult to find such things out, but also because the 'truth' tends to vary according to the positioning and perspectives of different actors: it is unfixed and variable. We shall return to these problems at the end of this chapter.

Effects

What are the social and cultural effects of development? This question is clearly closely linked to relative access. Rather than focusing on the distribution of benefits, however, it teases out different questions. By asking about the social effects of development, we are forced to consider the often complex social repercussions which may spill over into quite unexpected domains. Such questions are also vital in assessing projects or programmes which planners lacking in anthropological insight may not have originally considered to have any particular social implications, since these projects were primarily conceived of in technical terms.

Focusing upon social effects also demonstrates the highly complex nature of social change. People are embedded in a range of social, economic and political relationships which affect their access to property and labour, their decision-making power within their communities and households, their position in the division of labour and so on. Although anthropologists may not be able to predict exactly what the social effects of development will be, from what they may already know, and by asking the right questions, they are often far better equipped than most to make informed guesses. While the social effects of development must clearly be investigated during and after projects, through procedures of evaluation and appraisal, such questions also need to be posed at their inception. As we see below, the failure to do this has led to many grave mistakes.

Case 5: The Kariba Dam: the effects of resettlement[9]

Many large-scale projects which are designed to improve national infrastructure, and which are perceived as being solely technical, require the resettlement of large numbers of people. The building of roads, air-strips and dams to generate hydroelectric power provides classic examples. The social implications of these projects are often not fully comprehended until after they are under way and key questions, which might at least have limited the damage done to the groups that are forced to move, are not asked. The Kariba Dam is a classic example (see Scudder, 1980).

As Mair pointed out, when hydroelectric dams are built the displaced population is unlikely to benefit directly, for the electricity is usually intended for the inhabitants of distant cities (Mair, 1984: 110). The hardships caused for those who are forced to move can be reduced, however, if their social, economic and cultural circumstances are considered by administrators. In the Gwembe country (Zambia and Zimbabwe) where the Kariba Dam was built, there was insufficient consideration of these factors, even though many officials were deeply concerned for the people's welfare. In addition, a series of organisational mistakes were made. The worst of these was that, although the population was originally allowed to choose where they would relocate, a technical decision was taken to raise the level of the lake, resulting in the flooding of the area proposed for resettlement. This effectively destroyed any goodwill or confidence in the administrators that the relocatees might have had. While some villagers did move to sites they had chosen, at least 6000 were sent to the Lusitu Plateau, 160 kilometres away. Although the government had promised that water would be supplied, not only was the drilling machinery provided inadequate, but the water proved to be undrinkable, so that pipelines eventually had to bring water from the Zambezi River. In the time it took for these to be built, many people suffered from dysentery.

The people were moved to the area by truck. They were not allowed to return to Gwembe country. Since the administrators assumed they had no property, many valuable possessions were left behind or broken. The scheme also totally ignored the local organisation of work. Men were sent ahead to Lusitu to prepare the land and build houses in the very season when they would normally have been earning cash and clearing fields. Women were thus left behind to do all the agricultural work, while their men did tasks in Lusitu which traditionally women would have contributed to. On top of all this, compensation payments were inappropriate to customary property rights. Household heads were compensated for all the huts in their homestead, even though these were often built and owned by younger male relatives. A fixed sum of compensation was awarded to each individual, including children, but paid to the household head. Most of these shared out the money, but none shared equally; some young men claimed that they had to earn their share from their fathers by working for them first.

Although the problem of water supply in Lusitu was technical, most of the other problems relate directly to issues of an anthropological nature. Had key questions been asked before planning the move and the payment of compensation, many of the negative effects might have been avoided.

The list of questions is not of course comprehensive. It is also specific to Gwembe country. In different contexts, other issues may be important. For example, when squatter settlements are cleared, perhaps because a road is planned or simply because they are an 'eyesore', detailed questions

---

Effects: Key Questions

What is the nature of local power and hierarchy?
How is difference and inequality structured?
Are particular groups marginalised?
Do some groups monopolise political power and resources?

What is the nature of the household?
How is the household organised?
Who lives where?
How is decision-making power allocated within households?
How do these factors customarily change over time?

How are local property relations organised?
What goods are highly valued?
What access do different social groups or household members
have to property or other resources?
What are the usual patterns of inheritance?
How do these factors relate to the household development cycle?

How is work organised?
What are the main tasks done in the community, and during what
seasons?
Who does what work?
What is the importance of kinship roles or relations in the
allocation of labour?

*How suitable is the proposed relocation site, given the above economic,
social, and cultural factors?*

---

must be asked regarding people's relationship to the homes they live in,
tenancy arrangements and so on. There must also be safeguards to ensure
that opportunists do not claim property which less powerful individuals
occupy, or that 'household heads' are not given lump sums which may be
withheld from other members. It is vital that these questions are asked at
the planning stage, not after the project has already started.

Case 6: The Maasai Housing Project: technological change[10]

Since technology is usually produced, distributed, used and controlled by
different groups of people, changes in any of these areas are likely to have

knock-on effects on a range of social and economic relations. Different activities also involve varying amounts of power and status, according to each cultural context. Simply because some people produce a certain type of goods, for example, it cannot be assumed that they enjoy economic power, for they do not necessarily control its distribution and use. Likewise, people using a technology do not necessarily also control it. What implications does this have for projects involving technological change? The following example demonstrates that technological innovations and training in a housing project in Kenya have had various repercussions on local gender relations. These effects are by no means universal; rather, they depend upon the specific cultural context in which the project is taking place.

A report indicated that while some technological innovations in Kenya have had largely positive effects on women, for others the effects have been more mixed (ITDG, 1992). The Maasai Housing Project is a good example. Maasai women traditionally play a central role in the innovation, production, use and control of housing materials, but since the inception of the project their role in innovating new technologies has been reduced. In their place, men are becoming increasingly involved. Ironically, however, women's workload has increased.

The effects of the Maasai Housing Project must be understood in the wider context of Maasai life in Kajiado, Kenya. Although customarily associated with pastoralism, local Maasai have become increasingly settled. Alongside this more sedentary way of life, evidence indicates that women now shoulder greater burdens of work. For example, while men were traditionally responsible for livestock herding, women have now started herding, even though men still buy, sell and control the livestock. Most women work around 15 hours a day; lack of time is thus one of their largest problems. Other factors which prevent a greater share of decision-making power and access to resources for women are their lack of access to training and business opportunities, their lack of confidence, the threat of male violence, and their exclusion from decision-making and ownership.

The Maasai Housing Project was introduced to Kajiado District by the Arid and Semi-Arid Lands Programme (ASAL) in conjunction with a partner NGO. It started work in 1990, with the identification of 11 women's groups and the construction of a demonstration 'modern', three-roomed house. In 1991 women were invited to a workshop in which they expressed their own preferences regarding shape, size and interior of their ideal houses. The project then supervised the construction of five houses, three for rental and two for private use. A Maasai woman was employed as an extension worker, but the technical specialist and programme managers were men. A 1992 report suggested that women should be more central to the project: training courses should suit their time constraints, and housing designs should encompass their needs.

One problem was that the project's 'improved' houses took longer to build and thus added to women's work burden. While one woman reported that having a modern house gave her more status, most claimed that the greatest benefits were derived from technological improvements, rather than any social or political changes. Although it was hoped that one women's group would rent their house out while running a shop nearby in order to raise the money to provide it with better facilities, the group reported that this was not possible since they did not have the time or the money to run a shop. The house was thus left unoccupied.

Before the changes were introduced women were the main innovators and producers of housing; centrally, they also controlled the finished products. After the project, however, men were increasingly involved in innovation through their participation in training courses and in some aspects of construction (for example, carpentry). While women were still the main producers of housing, men had also started to distribute it. Combined with this, the values and statuses of each activity have also begun to change. Since modernity is highly valued in Kajiado, the distribution and control of modern houses leads to more status than that of traditional houses. This may be another reason why men are becoming increasingly involved. There is therefore a very real danger that men may increasingly control housing, while women will continue to do the bulk of the work and be the main users of the completed houses. Changes in the gender relations of house production may therefore also lead to changes in the gender relations of house design and control. Rather than benefiting from the project, women will be disempowered by it.

One way that these negative effects may be avoided is by ensuring that men are paid by women for their labour, thus giving them few rights over the finished product. Likewise, by improving traditional housing designs which are associated with female knowledge, male control of innovation might be reduced. It should also be remembered that the social relations of technology are not only culturally specific, they are also technologically specific. Housing among the Maasai is not an exclusively female domain. This means that men may choose to become involved in housing projects if they perceive that they will benefit from them. In contrast, other technologies are locally constructed as being exclusively female. For example, the production of stoves is seen by the Maasai as 'women's work'. Improved stove technology is therefore offered only to women by projects, without apparently discriminating against men. In this case, the new technology saves women time, rather than increasing their workload.

Control: Key Questions

How is local knowledge used, produced, distributed and controlled?
Who does what, and how is the work organised?
What is the relationship between these activities and decision-making power and status?

What are the constraints facing women?
How can project activities (training, group meetings and so on) fit most appropriately into women's tight work schedules?

How might the new houses be more appropriately designed?
Could the new designs be less, rather than more, labour-intensive?

What is the relationship between production, distribution and control?
Does the building and distribution of houses automatically lead to their control?

The Maasai Housing Project has not had wholly negative effects on local women. Indeed, great efforts have been made to recognise their productive role in house building and to enable them to participate in the design of new houses. The accompanying questions (see box) might, however, help 'fine tune' it.

Control

As the above case studies indicate, it is crucial to understand the dynamics of local societies if particular groups are not to be marginalised or further disadvantaged through development interventions. It would, however, be misleading to indicate that these issues are resolved solely through top-down planning. Indeed, this replicates dominant development discourses which presuppose that planning and policy-making simply need to be tweaked in particular directions to 'solve' the problems of development. Top-down planning is far from being the only solution. However well thought out development plans are, if they are designed and implemented by outsiders they are in continual danger of being unsustainable in the long term and of contributing to dependency; when funding ends, so does the project.

Unless people can take control of their own resources and agendas, development is thus caught in a vicious circle; by 'providing' for others, projects inherently encourage the dependency of recipients on outside funds and workers. Development discourses must therefore be challenged until they recognise that local people are active agents, and by changing their practices enable them to participate in project planning and implementation. In this section we indicate how development practice prevents people from taking control and how it might be changed from within. As in the rest of this chapter, we are confining our attention to planned change and assuming that, at some level, external donors are involved.

Working with local groups and institutions

Development plans often assume that the implementing agencies of a project or programme will come from outside the local community: there is a clear distinction between the 'givers' of a service or resource (development workers) and the 'receivers' (local people). Since developers are primarily interested in problems and solutions which are perceived in technological terms, local social structures tend to be seen as at best irrelevant and at worst an 'obstacle'. Indeed, outsiders often fail to recognise the degree to which communities have their own internal forms of organisation, decision-making and lobbying.

Unsurprisingly, however, projects are often most successful when they work through pre-existing social structures and institutions. For example, there may be pre-existing groups that are working to bring resources or services to their communities. These may take many different forms. For example, as we saw earlier, Latin American squatter settlements are often carefully planned by inhabitants, with local neighbourhood committees formed to develop the settlement. In other communities the group may have formed for a single purpose: gathering together to raise money for a school, a clinic or a place of worship, for example. Sports clubs are common forms of community-based groups, as are political parties. All of these vary from place to place; their suitability as implementers or partners for development work will depend on both their particular characteristics and those of the development plan.

Anthropologically minded advisers have an important role to play in contesting dominant discourses which ignore such groups. By finding out about them, representing their interests to planners, or enabling them to speak for themselves (for example, by arranging meetings or workshops), anthropologists in development can demonstrate a community's or group's potential for participation. Anthropological research and representation can also show that people are not passive 'recipients' but are accustomed to taking matters into their own hands.

Whether or not these groups become the basis for participation in a project is of course dependent upon a range of factors. The most important of these is probably the development agency's commitment to participation. It should also not be assumed that local groups wish to participate. As we shall see in the next chapter, much may depend upon the various meanings of participation being used. What is most important, however, is that such groups are asked what their interests are. They might decide that they need advice, training or extra resources. But they might just as easily wish to be left alone.

If there is a traditional system of communal decision-making, it may be easier and more expedient to use this as a participatory channel rather than creating new committees or institutions. If these institutions are dominated by a powerful elite, or particular groups are excluded, this may of course create problems, but simply to bypass local power-holders may cause greater difficulties in the long run. The work of Proshika, a Bangladeshi NGO, provided an example. While its projects were ultimately aimed at the local landless, organising them into groups and helping to raise their political consciousness in order to gain greater control of their situation, field workers often found it expedient to gain the trust of local elites and work through existing political structures. Where this was not done in the initial stages, the projects often met with fierce opposition.[11] In other cases existing committees or decision-makers might be linked to a new structure. In contexts where community decision-making is dominated by men, for instance, a separate woman's committee could be set up, feeding into the existing male-dominated one. If women are unused to being on committees or having a political voice, they may need particular support or training. Such projects cannot achieve miracles. Men may continue to dominate and women to have an unequal say in what takes place. But at least an opportunity for them to redefine their political roles has been provided.

Often there are NGOs already working within an area. Because these are smaller in scale than governmental agencies and are locally based, these often work far more successfully at the grassroots than bilateral aid projects[12] and are more experienced in participatory development. Increasingly, projects which aim to give beneficiaries greater control are attempting to work through NGOs already involved at the grassroots. Applied anthropologists may be asked to identify which local NGOs have the most participatory methodologies and which might be most able to carry out such work.

The following case study is an example of how project planning can build upon and strengthen pre-existing local groups and institutions in order to enable people to participate more fully in processes of change. As it indicates, development discourses are not homogeneously 'top-down'; they are both highly contested from within and liable to change over time.

Case 7: Labour welfare in tea plantations: enabling control[13]

A project to improve the quality of tea production in South Asia had been funded for several decades by a bilateral donor. Originally the project had been almost wholly technical, focusing on upgrading the quality of tea plants and productive techniques. While there was a labour welfare component, this concentrated on providing services for labourers within the plantations: improving their housing, providing tube-wells and health services.

By the late 1980s the labour welfare component began to be reappraised, not least because of ideological changes within the donor agency. Rather than simply providing services for labourers, policy-makers decided that the project should enable them to take greater control of resources; as much as possible, the project should provide a framework for the labourers to run their own project. This was politically highly controversial, for the plantations were owned by private individuals and companies, who wanted their labourers to be as passive as possible.

An anthropological consultant was hired to assess the viability of such plans by researching social structure and organisation among the labourers. What she found were high levels of pre-existing 'indigenous' organisation. Labourers lived in 'lines' of housing, within which foremen were appointed to oversee the maintenance of resources (such as tube-wells) and report problems to the estate management. Locally formed committees took responsibility for other decisions; for instance, those involving internal social affairs. Where resources (such as housing) had been provided by the plantation, there was a tendency to rely on the management of the estates to maintain them. Where labourers had built their own houses, however, they maintained them. Combined with this, in some estates female labourers were involved in managing credit and savings groups, an activity which appeared to have been initiated by the women themselves, rather than any outside agency. They also had their own indigenous healers and birth attendants, as well as the health services provided by the plantations.

Lastly, registered labourers were all members of the national trade union for tea workers. This had a long history of militancy. Local-level action – strikes, demonstrations and the garrotting of managers – regularly brought production to a halt in some estates. Each plantation therefore included union leaders, who had substantial experience in political organisation, lobbying and action. Many of the most forthright of these were women.

Thus while in some ways they had been forced into a passive role by the non-participatory allocation of services within the project, in other domains labourers were already actively taking control of affairs. Building upon this knowledge, project workers planned a new phase in the labour welfare component of the project. Local committees, based on the pre-existing organisation of the 'lines', would be set up. These would involve equal numbers of women and men; given the activism of some female labourers, it was reasonable to assume that this would not be too difficult. The committees would be based around the management and allocation of a 'social fund', to be provided through the project. It would be up to them how these funds were used. If they wanted to spend them on training, primary education or improved health services, they would decide.

Appropriate organisational structures

People are often excluded from participating in and ultimately controlling planned development because the organisational form it takes is inappropriate. Indeed, bureaucratic planning and administration are in many ways inherently anti-participatory, for they are deeply intolerant of alternative ways of perceiving and organising activities, time and information. Institutional procedures are therefore central ways in which development practices exclude supposed beneficiaries, even if superficially policy aims at 'participation'. These problems are not by definition insurmountable, but most bureaucracies will have to undergo major reorientations if their procedures are to be more open and flexible. Understanding the ways in which people are excluded by organisational structures and procedures means taking a step towards achieving this.

An example of the exclusive nature of planning procedures is the project planning framework, which some donors insist upon before providing funds. This involves an organisational chart in which planners specify project objectives, inputs, timings and the criteria they will use to measure successful output.[14] While this is undoubtedly a useful way of clarifying plans, the production of such a framework is also clearly much easier for administrators accustomed to particular ways of thinking and planning, and may require time-consuming training.

Project reports are another way in which administration and decision-making remain 'top-down'. Reports and other forms of documentation tend to be key to the formulation of policy within aid agencies, yet they are also often highly exclusiding to anyone from outside the institution. Reports are usually produced in very particular ways (for example, conventions such as listing recommendations at the beginning of the report, summarising information in appendices, keeping the text to a certain length, using particular bureaucratic phrasings and jargon). Those from outside the organisation who are not familiar with such conventions are thus effectively excluded from effective communication.

Projects which supposedly allow for local 'participation' are often planned in a way which makes such participation impossible. This is especially the case with projects which involve large-scale technical components, such as building. This tends to be planned around a rigid timetable and can usually be implemented relatively quickly. To set up the mechanisms for local participation in planning, however, usually takes far longer. Meanwhile those responsible for building are keen to progress as quickly as possible. These types of contradiction are extremely common, pointing to a larger problem in donor-led development: working with large budgets, which they are anxious to spend, donors and recipient governments are often reluctant to spend time 'fiddling around' with the complexities of setting up local committees and consulting communities about their plans. Instead, projects which absorb funds efficiently and are administratively relatively simple (building roads or dams) are preferred.

The timing of project activities may also be inappropriate. Again, this is the result of not consulting local people first. Meetings, for example, may be held at inconvenient times. Women may not be able to attend meetings or classes held at night. In other contexts women and men may not be able to attend those held in the day because of work demands. Once more these are issues which are best decided by the people involved. Anthropologists working in development should not take these decisions on behalf of beneficiaries, but wherever possible should ensure, at the very least, that plans involve careful consultation with them.

The location of project activities should also be considered, for they might be held in a place from which some people are excluded. In many Muslim societies women do not usually go into public places where there are many men. They may also be unable to travel to nearby towns to be trained, receive credit and so on, both for reasons of modesty and family honour and because they have domestic responsibilities throughout the day. Each context is of course different, but project activities are usually more accessible when they are decentralised.

Lastly, planners need to consider whether they are making appropriate demands on participants. As we know, men, and especially women, have to meet huge work demands in much of the world, yet this is often ill considered in the plans of outsiders. Projects which do not take these into consideration are therefore unlikely to gain much local support. A good example of this is income-generation projects which are highly labour-intensive. In the tea plantation project described above, an earlier plan in the labour welfare component of the project had been for income-generation activities for unregistered labourers, who often receive only very small incomes and, as a labour reserve, are not always in full employment. The problem, however, was that the plantations needed to have a continual supply of labour for times of high demand; if the unregistered labourers had an alternative source of income, the plantations would not have been so easily able to demand their work. The proposals were therefore blocked by the management.

Appropriate communication

People are often prevented from taking a more active role in development because it is conducted in cultural codes and languages which are alien to them. As we saw earlier, anthropological analyses of development discourse suggest that, by its very nature, it excludes people, disregards their knowledge and portrays them as 'ignorant', by upholding Western scientific rationality as the only paradigm for understanding and communication (Hobart, 1993).

While in the majority of cases this scientific rationality may provide solutions, this need not necessarily be the case. Again, the discourse is more heterogeneous and open to change than many commentators suggest. Anthropological knowledge has had an important role in promoting such concerns. It can also help to suggest more appropriate ways of getting messages across and enabling people to participate by using their own cultural idioms rather than those imposed from the outside. Again, this is not necessarily because anthropologists in development have 'expert' knowledge of a particular culture, but because they can insist at the planning stage that the advice of local people is sought.

Communication must be both appropriate and effective. The notion of appropriate communication may appear to be obvious, but it is extraordinary how often the local cultural and linguistic context is not considered in project planning. For example, in the early 1990s Katy Gardner sat in on a UNICEF training session for midwives in Orissa in east India, in which they were shown a training video made in the Punjab, several thousand kilometres away. The video was in Punjabi, and used traditional Punjabi implements and methods. Moreover, women sitting at the back of the small room could hardly see the video screen.

One simple way to communicate effectively is to use appropriate cultural forms. Community education projects often use traditional forms of entertainment to great effect. *Jatra*, or traditional travelling theatre in India, for example, has been used by community health projects to get across family planning messages. And in places where there is no, or very limited, electricity, communities may gather together to watch televisions powered by batteries. Again, this may provide a useful forum for showing films on public health, or other forms of community education.

But perhaps most importantly, planners must consider whether the message itself is appropriate. As anthropological analyses indicate, local knowledge is often based on assumptions that are quite different from those of 'rational' developmental knowledge (Pottier, 1993; Hobart, 1993). Training or education which disregards the ways in which people understand the world, and simply assumes that scientific or rational knowledge is accessible and useful, is therefore unlikely to be successful.

As we saw in Chapter 3, Richards argues that farming practices in West Africa can be understood as involving performance skills as well as detailed ecological and technical knowledge. Rather than skills being learned and 'set', farmers improvise their agricultural skills (P. Richards, 1993). Persuading farmers to adopt new seed varieties which have been developed in laboratory conditions because they are scientifically more advanced, or attempting to 'train' them in practices based on scientific understandings of agriculture therefore disregards the very nature of such farmers' knowledge and is unlikely to meet with much success. People understand events and ideas on their own terms. As long as development work involves the imposition of ideas and knowledge rather than being a dialogue, people are unlikely to be able to gain greater control of it, or voluntarily participate in it.

Conclusion

As the case studies cited in this chapter show, the more that is known about the dynamics and organisation of societies, at all levels, the more it is possible to ensure that particular groups are not excluded from or disadvantaged by planned change. Although one does not need to be an academic anthropologist to obtain this information, we suggest that understanding what questions to ask is primarily an anthropological skill. We are not suggesting that the insights and strategies discussed in this chapter should be confined to an elite of international anthropological consultants or 'experts'. Rather than certain individuals being the repositories of such knowledge, it is particular insights and methods which are important, and these are potentially accessible to everybody. Indeed, anthropological perspectives already inform much work being carried out by NGOs, and form the basis of methodologies such as participatory action research and participatory rural appraisal, which we discuss in the next chapter.

There is also no single way of gaining the sort of knowledge we have been discussing here. While traditional participant observation is certainly a possibility, such in-depth and time-consuming research is often not possible within the context of development work. The use of local consultants is nearly always preferable to hiring expatriates; local participants can also become 'indigenous anthropologists' – setting their own research agendas and answering questions on their own terms. Likewise, locally based NGOs often have extensive knowledge of local culture and social organisation (although this is not always the case).

The ease with which such information can be obtained should not be overestimated, however. Questions can be asked in any number of ways but there are no guarantees that the correct answers will be given, or even that there are 'correct' answers. To a certain extent, social realities always depend upon the subjective perspectives of those viewing the situation. Reality is also often highly contested; different interest groups will represent it in different ways (landlords and tenants, for example, are unlikely to agree about what the 'correct' level of rents should be). The ways in which outsiders are perceived may also influence how reality is represented to them. Researchers associated with aid agencies, for instance, may be seen as potential 'providers'. In these contexts it may be actively in people's interest to represent themselves more in terms of 'needs' than of self-sufficiency. In other contexts (for example, where researchers are associated with the government), local people may be extremely reticent to share information about landholdings, income and so forth.

Lastly, as we shall see in Chapter 5, while methodologies such as participatory rural appraisal offer alternatives to more top-down research, the danger is that they may easily be reduced to mechanistic gestures, a series of pre-specified activities which development workers carry out as quickly as possible, with little understanding of the rationale behind them, before getting started on the 'real' business of the project. Such dangers are exacerbated when projects are hemmed in by time-frameworks and targets.

In the next two chapters we build on these insights to ask how the new directions in Development practice, which, in the 1990s we suggested, were profoundly influenced by anthropological knowledge and questions, have fared. As we shall see, while some of the ideas that originally seemed to have the most potential have been co-opted by the dominant discourse, there are other areas and directions in which the possibilities for positive change and real alternatives remain strong. And where ideas *have* been co-opted, anthropological methods and our key questions of Control, Access and Effects have enabled anthropological critiques to be as powerful as ever. This work of engaged critique is perhaps even more pressing, for claims of 'partnership', 'participation' and 'empowerment' and so on made by corporate bosses, local government officials and virtually every NGO seeking donor funding, may be deeply misleading.

5

When Good Ideas Turn Bad: The Dominant Discourse Bites Back

By probing the nature of power and asking who benefits from planned and unplanned change, who controls it and what the effects are on different groups, anthropologists have long critiqued many of the assumptions of mainstream development discourse. In the first edition of this book, replicated in the last chapter, we argued that in planning or evaluating the work of development, the questions of *access*, *control* and *effects* required urgent attention. Yet if ignoring them had led to the failure of many schemes of improvement and we remained sceptical about much conventional development practice, we also found reasons to be cheerful. Indeed, the book showed how anthropological ideas and methods were core to some of the most exciting approaches emerging in development in the early 1990s. Along with feminism, anthropology had informed a number of seemingly radical new methods and ideas which, by the mid 1990s, were starting to make headway into the mainstream. These included perspectives on local knowledge and 'farmer-first' approaches that contributed to Robert Chambers' pioneering work on participatory approaches (Cornwall and Scoones, 2011), radical analyses of power and empowerment (James, 1999), and feminist anthropology's huge contribution to what was to become a sub-field in development studies, namely gender and development (GAD; Staudt, 1998).

On the surface some of these gains seem to have been sustained over the intervening years. Indeed, innovations which 20 years ago were radical outliers have become accepted into the fold in a way in which their original advocates might have found amazing. If participatory rural appraisal (PRA) was a brave new idea in 1993, by 2014 it is as conventional and everyday as a log-frame or impact assessment. Meanwhile, as Rosalind Eyben (2004) describes, while in the 1980s feminist perspectives on development were treated with surprise and suspicion within what was then the ODA, by the late 1990s they were increasingly mainstream in the newly formed DFID. 'Empowerment' seems to have had a similar success, making routine appearances in policy documents speeches and websites.

Yet appearances can be deceptive. As seemingly radical practices have been taken on board, mainstream development has neutralised or turned them into something very different. Rather than being accepted in their original guise, anthropological research shows how they have been *made acceptable* to the dominant discourse. Worse, in some contexts they have become window-dressing, feel-good terms employed to make agencies or businesses look good without engagement with their original meanings or implications: 'green wash' of the most depressing kind (see Kirsch [2010a, 2010b] on sustainability as an oil industry oxymoron).

Was our original optimism misplaced? In 1996 we cited Escobar, who had written that : 'Although the discourse has gone through a series of structural changes, the architecture of the discursive formation laid down in the period 1945–55 has remained unchanged, allowing the discourse to adapt to new conditions' (1995: 42; cited in Gardner and Lewis, 1996: 103). At the time, we disagreed, arguing that rather than being as monolithic as Escobar and the other deconstructionists seemed to claim (see Olivier de Sardan, 2005), development was heterogeneous and able to change from within. In this and the last chapter of the book we return to these questions, paying particular attention to the fate of what back in the early 1990s seemed to be among development's brightest hopes: 'empowerment', gender and development, and participatory research techniques. By drawing upon the conceptual apparatus of discursive analysis some of the ways in which these ideas have spun free from their radical roots are revealed.

This does not mean that the anti-politics of development is always successful or that there have not been other shifts in ways of seeing and doing. If development has turned participation into a box-ticking exercise and 'empowerment' into a photo opportunity there are other areas where ideas and innovations have spun free from the de-politicisation and standardisation of project and policy. Indeed, simply because some agencies or organisations are misusing or abusing terms or practices does not mean that in other contexts they have lost their bite. And, as we shall see, rather than being narrowly corralled as a set of activities aimed at assisting or improving the global South, 'development' has become something far wider, dissolving boundaries between North and South, and breaking free from the institutional constraints of Development World. We end the chapter with some brief examples of the role of anthropology in these new directions, as well as the continued importance of critique.

When good ideas go bad: the rise and fall of politicised development

One way of tracing the ways in which ideas become co-opted and dumbed down is to focus in detail upon on the way the discourse actually works, including the genealogy of its actual language.[1] In their edited volume *Deconstructing Development Discourse: Buzzwords and Fuzzwords* (2010) Andrea Cornwall and Deborah Eade wittily describe how 'Development-speak' involves a lingua-franca of phrases and words in which those working within Development World must learn to be fluent in order to be accepted into the club. This language has hegemonic properties; if Southern researchers or practitioners are to break into the international development market they are forced to use a lexicon already determined by the ruling elite. The language is not static, however. Instead, Development-speak, which is always shaped by those in control, involves a constant process of:

> burying out-moded jargon, authorising new terminology and permissible slippage, and indeed generating a constant supply of must-use terms and catchphrases.... The extraordinary thing about Development-speak is that it is simultaneously descriptive and normative, concrete and yet aspirational, intuitive and yet clunkingly pedestrian, capable of expressing the most deeply held convictions or being simply 'full of sound and fury, signifying nothing'. This very elasticity makes it almost the ideal post-modern medium, even as it embodies a modernising agenda. (Eade, 2010: viii–ix)

While Developmentspeak buzzwords allow ideas to circulate rapidly, their fuzziness means that they can mean all things to all people, carrying the excitement, newness and even edginess of the buzz, yet remaining, at heart, uncontroversial and unproblematic because of the fuzz. Were it not for the political implications, we might compare it to the ways in which street styles are spotted by the fashion industry, mass produced and, within a short while, rendered dull. Like mass-produced fashion, fuzziness allows the terms to reach a global market. Terms such as 'community', 'social protection' or 'sustainability', for example, carry multiple agendas, allowing them to circulate between agencies and groups under the guise of a shared meaning which may have existed at the outset but has slipped over time, and always towards the definitions of those in control. It is this, not the terms' mass appeal, which is crucial. Since those in control do not promote radical change or political critique, their promotion of buzz- and fuzzwords is inherently conservative, pushing for the least challenging interpretations and practices. As Andrea Cornwall puts it with reference to community participation, by the 2000s 'doing it for yourself' has become 'doing it *by yourself*' (Cornwall, 2010).

One way in which ideas become depoliticised is a result of their fuzziness; their very vagueness and ambiguity allows them to mean different things to different people, which allows them to become attached to widely disparate practices without contestation or debate (Cornwall, 2010). Thus while the term 'empowerment' has radical meanings for some, involving consciousness-raising, the critique of existing structures and, its inevitable end product, political action in order to change those structures, for others it simply means 'helping people to help themselves'. The anti-politics of the latter statement is never about politicised structural change; instead, it tends to be linked to practices which perpetuate or introduce neoliberal tenets of individualism, entrepreneurship and so on. Income-generation schemes which tout the power of markets and aim to 'capture' the commercial potential of the bottom billion while training individuals to be sales persons for corporate interests such as Avon are a prime example (Dolan and Johnstone-Louis, 2011).

To throw more light on these processes let us turn to three key areas within development: empowerment, gender and development and participation. As we argued in 1996, all share close synergies with anthropological approaches, either methodologically or because the questions that anthropologists ask concerning access, control and effects, and the nuanced ways in which anthropologists theorise power, have contributed to their gestation and growing importance within development work. As we shall see, in many ways all three have lost their radical bite.

Let us start with 'empowerment'. As with each of the following ideas, we begin with some background history, drawn from our 1996 edition:

Empowerment

Empowerment has been described as being 'nurturing, liberating, even energising to the unaffluent and the unpowerful' (Black, 1991: 21). This version of empowerment is in part drawn from the ideas of the Brazilian educationalist Paulo Freire, based on the need to stimulate and support people's abilities to understand, question and resist the structural reasons for their poverty through learning, organisation and action. Linking disempowerment with illiteracy, and illiteracy with particular modes of learning and ways of knowing, Freire (1968) argued for education as revolutionary consciousness-raising, a means by which the oppressed could overcome their conditions. To this extent, processes of empowerment necessarily involve a critique of the structures of oppression. Literacy programmes promoted by some Bangladeshi NGOs in the 1970s and 1980s, for instance, used teaching materials which both gave learners the means to become literate and heightened their political consciousness via discussions of exploitative money-lending arrangements or land grabbing. The aim was to create a politicised peasantry, capable of challenging the wider structures which created poverty.

In a related approach, John Friedmann's (1992) analysis of the politics of alternative development outlined a theory of poverty which viewed it not simply as the absence of material or other resources but as a form of social, political and psychological disempowerment which must be challenged. In this view, whole sections of the population – landless rural workers, subsistence peasants and shanty town inhabitants, for example – were systematically excluded from participation in the development process. Friedmann therefore made empowerment the central aim in his discussion of the politics of 'alternative development':

> The empowerment approach, which is fundamental to an alternative development, places the emphasis on autonomy in the decision-making of territorially organized communities, local self-reliance (but not autarky), direct (participatory) democracy, and experiential social learning. Its starting point is the locality, because civil society is most readily mobilized around local issues. (1992: vii)

Friedmann saw the need for alternative development models to acknowledge the rights and established needs of citizen households and individuals, which involved a political struggle for empowerment and against structural constraints. For example, the NGO Proshika's work in Bangladesh in the 1980s and 1990s included group formation, in which landless people took action in pursuit of their rights against locally powerful individuals. In one example documented in a collection of case studies from Bangladesh, groups of landless people in Gazipur district successfully organised a public boycott of a local landowner who was engaged in stealing public agricultural land by securing false land-title documents. The landowner had no access to public transport or hired labour and suffered public humiliation, and the group members who had lost rightful access to the land won the legal case against him in the courts (Kramsjo and Wood, 1992: 63).

Yet, even at the time of writing the first edition of this book in 1996, empowerment had become a frequently degraded term in mainstream development. Rahnema (1992: 123) saw the term simply as providing development discourse with a new form of legitimation and convincing people 'not only that economic and state authorities are the real power, but that they are within everyone's reach, provided everyone is ready to participate fully in the development design'.

In some countries, governments talk glibly of empowerment of the poor in their development plans, having stripped the term of any real meaning. In other planning documents there is an assumption that empowerment can be achieved simply by providing credit to low-income people. As Korten (1990) notes, it is not really possible for one person to 'empower' another: people can only empower themselves. Korten argues that this requires a process of 'mutual empowerment' in a group setting, often with outsiders as facilitators. The danger of creating dependent groups, well-versed in the rhetoric of consciousness-raising but remaining essentially unchanged by the experience, was observed in Bangladesh by the mid 1980s (Hashemi, 1989).

By the twenty-first century it was not unusual to hear some very un-radical people talking blithely of 'empowerment', which had by now also entered the world of management speak (see James, 1999, 'empowering ambiguities'). Take the 'Community Engagement' programmes of the multinational energy corporation Chevron in Bangladesh (which we will elaborate on in the next chapter). In 2012, the term 'empowerment' was being used without irony or embarrassment by an official tasked with promoting the external relations of the company. As he put it, with reference to the financing of credit and income-generation programmes in villages surrounding the company's gas field in Bibiyana, which had resulted in loss of land and livelihoods and caused significant protest only a few years earlier: 'Our aim has always been to *empower* people' (Gardner, 2012). Ensuing discussions with company officials, however, revealed that local views of the company's presence (which were largely negative) were dismissed by staff as 'hearsay', and problems of 'elite capture' accepted as standard 'for projects of this type'. Meanwhile, project success was celebrated in the company's publicity material via photographs of 'handing over ceremonies' which largely involved Chevron officials making development gifts to individual recipients, many of whom had been chosen via pre-existing networks of patronage which the programmes did not challenge (Gardner, 2012).

Here, then, we have the use of terms such as 'empowerment' which allow a corporation to appear to be 'doing good' while involving remarkably little of what most left-leaning anthropologists would consider empowering. In this instance, ideas which arose from progressive and leftist roots are used by multinational corporations to further corporate interests, stripped of their original intent and turned into a series of exercises or performances of project success (see Mosse, 2005). As Srilatha Batliwala recounts of the Indian context, while the term started off in the 1970s as overtly political, focused upon power relations and, in the women's movement, women's unequal access to resources, by the 1990s it was losing its political edge, becoming 'demoted' to the domain of 'self-help' groups or even, in some instances, linked to Hindu fundamentalism. As she puts it:

> empowerment was hijacked, in the 1990s, into increasingly bizarre locations, converted from a collective to an individualistic process, and skilfully co-opted by conservative and even reactionary political ideologies in pursuit of their agenda of divesting 'big government' (for which read: the welfare state) of its purported power and control by 'empowering' communities to look after their own affairs. (Batliwala, 2010: 112; see also Batliwala and Dhanraj, 2007)

If the slippery nature of words and their susceptibility to falling into the hands of the wrong people is partly to blame for the de-politicisation of what were originally radical ideas, the simplification of complex concepts into feel-good sound-bites or 'easy on the eye' images is another. This process is compellingly described by the contributors to Cornwall et al.'s (2007) revisiting of gender and development's 'myths and fables'. Let us turn to our second 'New Direction' of 1996: gender and development.

Gender and development

As we showed in 1996, during the 1970s and into the 1980s gender relations were increasingly recognised as central in determining people's access to resources and the ways in which they experienced development. This was largely a result of the work of feminist anthropologists and, we argued, one of the main ways in which anthropology had directly affected development policy. In what follows we sketch some of the background to this, before turning to recent work which casts doubt on some of the results.

A major step towards official acceptance of the need to consider the relationship between development and gender more carefully came in the guise of the UN Decade for Women (1975–85). During this period there were important changes in the ways both policy-makers and academics approached gender. Whereas previously both groups had tended to concentrate on 'women' and their domestic reproductive roles, policy by the mid 1980s was increasingly emphasising women's employment, income-generation capacities and so on, rather than the provision of welfare services for them.

The UN Decade marked what appeared at first to be a growing institutional commitment to women's issues, although the rationale behind this varied. Prompted partly by the work of writers such as Boserup, and also as a reflection of the successes of feminism in the North, which had enabled a few women to reach managerial positions within aid agencies and had pushed feminist issues on to the political agenda, by the early 1980s many development agencies had determined to 'do something' for women. For example, in Sweden in the 1970s and 1980s parliament was subject to successful lobbying pressure by Swedish women's organisations for official aid to address specifically women's needs and this became reflected in SIDA's[2] programmes. USAID also rapidly adopted the new phrase 'women in development', with the establishment of an Office of Women in Development. Although the meanings of 'women in development (WID) are far from fixed, USAID seemed to use it in terms of the potential contribution women could make to the development effort, as a so far untapped resource. Many other institutions followed suit, setting up WID offices or, like the British ODA (now DFID) building a commitment to women into official policy. Indeed, it is now commonplace for government ministries, NGOs and multilateral agencies to pay lip service (if nothing else) to the aims of WID, and some donors insist on a WID component in project proposals before they consider funding.

The WID approach, however, tended to focus only on women in isolation, rather than the social, cultural and political relations of which they are a part. As feminist anthropologists pointed out, it was gender and not sex which was at issue. This led to a shift towards GAD, which turned attention away from women as an isolated category to the wider relations of which they were a part. It should, however, be noted that the terms were often used interchangeably, and policies all too frequently focused attention only on women. Indeed, despite the energy and resources directed at gender issues, by the mid 1990s WID/GAD still frequently remained an 'add-on' to mainstream policy (Moser, 1993: 4).

WID/GAD approaches were far from homogeneous. In her account of WID projects, Caroline Moser outlined five main approaches, each associated with a distinct developmental philosophy (1989: 1799–825). While we must beware of over-schematising (for example, policies and projects often involve a variety of assumptions and approaches), this clearly indicates the range of responses to gender issues within development practice. A *'welfare'* type project, for example, is linked to charitable notions of 'doing good' for women and children, and involves the top-down provision of services and goods for beneficiaries, without demanding any return on their behalf. While this approach was common in the 1960s and early 1970s, with the growing influence of feminism as the 1970s unfolded, notions of *'equity'* increasingly gained sway in some development circles. These aimed at boosting the rights and power of women within developing countries, again usually through top-down changes in governmental policy, state intervention and so on.

Another approach which gained popularity in the 1970s and 1980s was *'anti-poverty'*, in which poverty was recognised as women's main problem. This was closely allied to the *'basic needs'* movement, which had taken off during the 1970s. Solutions included income-generation projects, skills generation and so on. These strategies were often identical to those advocated by the *'efficiency approach'*, but their underlying philosophy was fundamentally different. Efficiency was central to much developmental philosophy during the 1980s, in line with the dominant political ideologies of the time. Accordingly, women were targets of development projects only because the centrality of their productive contribution was recognised. If projects aimed to improve recipients' wellbeing, rather than being based in notions of welfare or universal human rights, the underlying philosophy was that this would, in turn, increase their efficiency in the productive process and thus add to capitalist growth.

All of these approaches assumed that change is initiated first and foremost from the outside, through donor-led policies and planning. As well as being fundamentally 'top-down', they were accused of being ethnocentric. Many of their fiercest critics were Southern women, who argued that discourses of WID/GAD reflected the preoccupations and assumptions of Western feminists rather than the women they purported to be representing and assisting. Indeed, by homogenising all 'Third World' women (in concepts such as 'female-headed households', or in policies which treated the interests of women in vastly different cultural, economic and political contexts as the same) and treating them as victims in dire need of policies which alter their status, these approaches fed into colonial stereotypes and categories (see, in particular, Mohanty, 1988). By treating them as 'victims' of their culture, it was argued, they negated and undermine the *agency* of Southern women (White, 1992: 15–22).

Another criticism made of WID/GAD approaches was that they made ethnocentric assumptions regarding the content of relations between men and women in different societies, seeing only exploitation, subordination and conflict, whereas the women concerned might put more stress on cooperation and the importance of familial bonds (Barrios de la Chungara, 1983). Lastly, WID/GAD approaches were accused of ignoring the true underlying causes of Southern women's subordination and poverty, which are more to do with colonial and postcolonial exploitation and inequality than the cultural construction of gender within their particular societies (Sen and Grown, 1987).

Having said this, many of the institutional and policy changes regarding GAD were welcomed by feminist anthropologists in the mid 1990s. But even then, in their early days, the policies illustrated the capacity of radical concepts to be neutralised within development discourse. At worst, the effect of WID/GAD approaches in development was to transform what were in reality complex and nuanced conceptual tools and insights into overly simplified categories and phrases, which nonetheless were made central to policy (such as 'women-headed households') but effectively stripped of their radical implications.

By the early 2000s it was clear that the gap between feminist aspirations for social transformation and actual gains was significant, despite what appeared to be huge inroads over the 1980s and 1990s. The period had witnessed the opening up of important spaces for feminist engagement with development, including the official adoption of gender mainstreaming by the UN at the 1995 World Conference on Women in Bejiing, the apparently successful integration of women into much development policy and practice, widespread gender training in institutions such as DFID, hugely increased funding for projects focused on women, and so on. Yet as the authors of *Gender Myths and Feminist Fables* recount, the success of feminist agendas has been much more elusive than many originally supposed:

> arguments made by feminist researchers have become denatured and depoliticised when taken up by development institutions. For many, what were once critical insights, the results of detailed research, have now become 'gender myths': essentialisms and generalisations, simplifying frameworks and simplistic slogans. (Cornwall et al., 2007: 1)

The processes by which this happens are important, for they are integral to much of what constitutes development. Again our attention is focused on the workings of the discourse, in which writing policy documents – a particular genre – is a major part of the job (Karlsson, 2013). As Cornwall et al. (2007) recount, policy documents require clear stories with obvious villains, victims and saviours. As Roe (1991) shows, such stories are not necessarily untrue, but they are shaped in a particular way and tend to encourage generalisations or simplifications. They also follow particular narrative arcs: for example, that women are victims and need saving by development policies or projects. Without malicious intent, the effect is that the complexities of feminist research are lost, replaced by myths, framings and simplifications: the sex worker as a victim, for example, women-headed households as poverty stricken, or African women as pregnant, illiterate and poor. These myths can be difficult to dislodge and may perpetuate negative stereotypes or silence other groups, such as middle-class African women, for whom issues such as HIV/AIDs or domestic violence remain salient, but who are never even imagined, let alone targeted, in development interventions (Win, 2007).

One of the challenges facing work which aims to address gender inequalities is that success is difficult to measure. Institutionalised development has a habit of relying on metrics or 'indicators' to evaluate whether or not a project has worked, meaning that potentially complex outcomes are reduced to statistics. While this is not necessarily a problem if what is to be counted are the numbers of vaccinations given, slab latrines installed or new seeds distributed, the process becomes difficult, if not ludicrous, when levels of empowerment are being calibrated. How are we to know if women have more or less power within their homes, communities and relationships as a result of, say, education, employment or access to credit? How, indeed, can we trace shifts in gender norms to particular causes, when, as anthropologists have always shown, cultures, ideologies and practices are intertwined with political and economic structures in such complex ways? While this is has been widely debated within academia (cf. Kabeer, 1994, 2002) the use of simplistic indicators can hide all manner of obfuscation and misunderstandings. An increase in women's political participation *might* be indicated by more women sitting on local councils or other public decision-making bodies, and this *might* result from micro-finance programmes. Then again, the women sitting on the councils *might* be there in name only, as research in Bangladesh has found, in order to satisfy funders that the project is successful. As Mosse (2005) has shown, a great deal of effort is put into performances of success; these performances are made all the easier when indicators are simple and measureable.

Even if women are not formally participating in public bodies, or, as other research has indicated, the credit they receive from micro-finance groups goes to their male relatives, does this mean that they have received no benefits and the project has failed? (cf. Karim, 2011; Goetz and Sen, 1996; D'Espallier et al., 2009; Pitt and Khandker, 1998; Roodman and Morduch, 2014). The answers would involve listening carefully to women's accounts of how the project has affected their lives and understanding the wider cultural context, as well as observing actual changes in behaviour. It is not that change can never be measured, more that the measurements require caution. Is an increase in reported sexual assaults against women a result of increased misogyny, for example, or of women's greater confidence that if they report an assault the police will assist them?

Institutionally, there is often a tendency for policy and the practices which ensue to simplify what were intended as more complex feminist agendas, especially when agencies are multi-layered or international donors are working with in-country organisations. It's easy for the message to get lost or diluted; policies end up focusing on women's gender *needs* (for example, reproductive health) rather than their *strategic goals* (for example, policies which promote economic or political participation; see Moser, 1993). As Hilary Standing has described, the politics can be troubling, with national agencies being ticked off by international donors for promoting needs but not strategic goals. Foot dragging and slippage of policy goals may result in part from the nature of large bureaucracies, which tend to move slowly and in which it is difficult to introduce change. But they also belie a larger problem facing GAD: the inevitable resistance to what are essentially feminist challenges to patriarchal power. This may exist within institutions (see, for example, Mukhopadhyay, 2004), but the obstacles are also very much outside them: the violent attacks on NGOs in Bangladesh in 1994 (see Lewis, 2011a) or, indeed, on girls attending schools in northern Pakistan and Nigeria in 2013–14, are a vivid reminder of just what such programmes might be up against.

As a buzzword, 'gender' has therefore proven easy to place on the development agenda, but, once there, it has been difficult to make the concept work towards progressive change. Cornwall concludes that the term has met a sorry fate:

> In more recent times, it has fallen from favour and has a jaded, dated feel to it. Diluted, denatured, depoliticised, included everywhere as an afterthought, 'gender' may have become something everyone who works for an aid organisation knows that they are supposed to do something about. But quite what, and what would happen if they carried on ignoring it, is rarely pungent or urgent enough to distract the attention of many development bureaucrats and practitioners from business as usual. (Cornwall et al., 2007: 6)

Yet while there is little doubt that much of the 'feminist intent' of GAD has been lost by its mainstreaming (Cornwall et al., 2007), the picture is not wholly negative. In the next chapter we discuss innovative approaches which side step many of the problems faced by more conventional programmes. Let us turn to our last 1996 'New Direction': participation.

Participation

If gender has become a moribund buzzword (a dead word?) which few development officials know quite what do with, participation has met the opposite fate. Transformed into a series of do-it-by-numbers practices and exercises variously packaged as 'PRA' (participatory rural appraisal) or PLA (participatory learning and action), it has proven remarkably resilient to development's fickle appetite for new ideas and methods. Indeed, there are few projects and schemes today which do not involve some element of 'participation', or at least mention the term in their documentation. But is this because it 'works', because it's relatively easy to carry out successfully, or because it's easy to give the *impression* of it having been carried out successfully? What is it anyway, and what does it mean for 'it' to have worked?

Let us recall the history (see Cornwall and Pratt, 2011). Participatory approaches in development can be traced to several distinctive roots. In the first, participatory research techniques draw upon the ideas of 'participatory action research' (PAR), a method linked to Freire's notions of consciousness-raising empowerment. As Rahman described in 1993, PAR involved the poor and marginalised researching the causes of their own oppression, a collective research process which would lead to action that would precipitate social, economic and political transformation; it was this and not 'development' that was its main aim (Rahman, 1993). PAR was directly and radically political. Over the 1970s and 1980s, for example, PAR was associated with revolutionary movements in places such as Guatemala and Nicaragua, which Freire visited (Leal, 2010: 92).

In the second strand of meanings, the aim of participatory methods was to improve development practice. Drawing upon anthropological work on indigenous or farmers' knowledge, which showed how, in failing to take local perspectives into account and assuming that scientific knowledge would lead to positive change while local knowledge or culture was an obstacle to progress, many schemes were bound to fail, Robert Chambers (1992) set forth a compelling argument for a research methodology within development which would prioritise local viewpoints and knowledge. Chambers argued that, rather than basing their understanding of rural communities on the view from their jeeps, development experts should spend time listening; indeed, development interventions should be based around local analyses of the problems and local solutions via a method which became first known as 'participatory rural appraisal'. This involved using culturally appropriate techniques to gather community knowledge: developers and community members would sit on the ground, perhaps using sticks to draw maps in the sand or using beans for counting. As Cornwall and Pratt note though, this image is somewhat overdrawn; rather, participatory methods involve groups using their creativity to analyse situations and problems (Cornwall and Pratt, 2011: 264).

The method was taken up with huge enthusiasm. Not only did PRA *make sense* (who could refute the premise that talking to local people and understanding matters from their point of view might be a good idea?) but it felt and looked good, too. As with GAD, it was also easy for the radical, consciousness-raising elements of the approaches to be quietly forgotten and put aside.

The timing was perfect. By the early 1990s discourses of rights were becoming widely accepted; postcolonial theory had challenged the meta-narratives of modernity, social movements which stressed the identities and specific concerns of minorities were gathering force. As we reported in 1996 and later in 2000, while development faced accusations of top-down imperialism, there were moves from some quarters within to make real changes to the way that development work was done: the funding of NGOs, promotion of social development advisers and stress on local partnerships were part of this. 'Participation', as a set of methodological tools, was ideally placed to bridge the gap between these more progressive agendas and the inherently conservative nature of mainstream development institutions. While Chambers and others may have been advocating a genuine foregrounding of local knowledge, bottom-up approaches and the participation of communities in the planning of development schemes based on flexible and evolving participatory approaches, in many contexts 'PRA' became a standardised toolkit of methods ideally suited to the anti-politics of institutionalised development. Trainings, workshops and PRA materials followed, all of which were easy to count and to tick off as another outcome or indicator that good work had been successfully carried out. Stakeholder workshops, for example, follow standardised management models, allowing policy to be put into practice and social spaces for project personnel to be opened up in a remarkably uniform way (Green, 2003).

The implementation of PRA proved to be tricky, too. In *Cultivating Development* David Mosse (2005) describes how, in the Indian irrigation project he was working on in the mid 1990s, development consultants set about PRA with initial enthusiasm, only to learn the hard way how difficult it was to get 'the community' to come together and share their perspectives. Indeed, as many anthropologists of development and practitioners of participation have pointed out, PRA involves a number of deeply problematic assumptions: that 'the community' is homogeneous, that indigenous or local knowledge can be accessed via participatory exercises or is held uniformly by all people, or that PRA techniques can include and give equal value to the views and knowledge of the less powerful: groups such as women, children or minorities, who might not find a voice or be silenced during PRA events, if they were even to attend (see Cornwall, 2003). Others questioned the dualistic thinking implied in

PRA approaches: that knowledge was local or indigenous and therefore opposed to, or strikingly different from the 'scientific' knowledge of development practitioners (Scoones and Thompson, 1994). Increasingly, practitioners raised concerns over the ritualistic nature of PRA, suggesting that, like so much in development, it may have quite unintended consequences, not producing 'participation' or local knowledge but its shadow: the *impression* of participation, allowing developers to believe that their schemes are inclusive and participatory and to thus present them as successful. Over the 1990s, PRA thus became associated with myriad tales of abuse (Cornwall and Pratt, 2011: 264–65). Mosse, for example, shows how, in the Indian project where he worked, participants adapted representations of their interests according to what they believed the project might offer; it was not that PRA failed in this context. Rather, it became an important system of representation, allowing different groups to mediate their interests (Mosse, 2001). As Cooke and Kothari (2001: 7) conclude in their critique of PRA, while there are technical challenges to PRA which might be fixed, the more substantial challenge is 'not with the methodology and techniques, but with the politics of the discourse … with what participatory development does as much as what it does not do'. Once again, we are returned to what we might think of as 'The Fall': the co-option and ruin of good ideas by what seems to be hegemonic development discourse.

To understand how and why this has happened, we must turn to the wider historical context. As Leal (2010) describes, the enthusiastic uptake of PRA methods by mainstream development institutions coincided with drastic structural adjustment programmes in the 1990s which were far from participatory or empowering for those affected by them. Agencies such as the World Bank took on the language of participation and empowerment with aplomb, setting out strategies which justified curtailing the power of states which, the World Bank's documents suggested, had taken resources from 'the people'. Rather than returning such states into the hands of the people, however, the World Bank's solution was to introduce neoliberal policy and the market as a form of 'empowerment'. Leal writes of the participatory 'products' which the World Bank used to promote their version of participation over the 1990s, using particular discourses and techniques to make them appear to be caring and democratic. These include the *Participatory Source Book* in 1996, the *Voices of the Poor* study in 2000, Poverty Reduction Strategy Papers (PRSPs) and participatory poverty assessments (Leal, 2010). As Leal writes, the Declaration of the Millennium Development Goals is: 'peppered with buzzwords such as "sustainability", "participation", "empowerment", "equality" and "democracy" but ... makes no reference to what might be the forces that produce and perpetuate poverty' (Leal, 2010: 94). The result is that the language and techniques of participation effectively mask underlying political processes which uphold the status quo, promoting the market and the World Bank's continued grip on poor countries via the technification of social and political problems. In this reading, participation 'greases the wheels' of the anti-politics of development (Williams, 2004, cited in Leal, 2010: 95; see also Cooke and Kothari, 2001). In her research on workshops and 'micro-entrepreneurs' in Cairo, Julia Elyachar (2002, 2005) terms such policies 'anti-development development', revealing how what were once seen as 'obstacles' to development – the cultural capital of the poor, involving informal systems of exchange – are now celebrated by the World Bank, which uses the idea of self-reliance as a justification for the neoliberal withdrawal of state and donor support.

*   *   *

If that was the bad news, how about the good? While it is important to understand how ideas and practices change over time and in some contexts lose their political bite, this does not mean that all is lost. Rather, the basic ideas that, in the mid 1990s, we argued had the potential to effect positive change within development – the need to involve people in planning their own futures, understandings of difference, whether based in gender, class, generation or other signifiers, and close attention to the dynamics of power – remain wholly relevant. Indeed, in a world largely governed by finance and the apparently unfailing belief of governments and donors in 'the market' and/or economic growth, the focus on access, power and effects remain more important than ever. These are clearly political questions, but they are also questions which anthropologists are supremely well-qualified to ask, and to answer, both via their methodology and their theoretical perspectives. We should also recall that critique is in itself political and has the capacity to lead to change (Hale, 2006). For example, sustained ethnographic research, underpinned by the theoretical positions laid out in Chapter 3, has produced critiques of taken-for-granted orthodoxies and 'anti-development development' (Elyachar, 2005). This work of critique plays an important role in undermining homilies, attacking feel-good spin and asserting new agendas, and is, we suggest, one of the discipline's greatest strengths. In what remains of this chapter, let us turn briefly to some examples of the continued relationship between research, critique and action.

Ethnography, participation and empowerment, round two

If, in some contexts, good ideas turn bad, can they be saved? Indeed, have they 'gone bad' all over, or just in some places? Before we rush to condemn participation out of hand, or assume that because a depoliticised development has largely failed to deliver women's empowerment as an 'outcome' empowerment never takes place, let us consider the following points: first, merely because a method or approach has been co-opted and neutralised in some quarters does not mean that in other contexts it has nothing to offer. If the 'participation' of the World Bank is less than radical, other groups, organisations or movements might use the approach with more meaningful results. As Cornwall and Pratt (2011: 264) argue, PRA involves a vast array of practices; merely because it is not done properly in some contexts does not invalidate its aim: participatory development, in which the people to be affected by development schemes participate in the formulation of projects and policies. For many practitioners these aims remain core. As Cornwall and Pratt put it: '"Participation" has come to embrace a plurality of normative and ideological positions, from communitarian impulses towards neo-liberal "do it yourself," "community-driven" strategies, to radical democratic notions of expanding the boundaries of the political, all of which have histories ... and all of which imply different purposes to which technologies like PRA may be put.'

As this implies, ideas are not static. Instead, they are continually evolving, shaped according to different circumstances and changing according to time and place. Thus participatory approaches are constantly subject to change and revision, evolving into approaches to effect action rather than simply gather knowledge which the developers take away (cf. Chambers, 1992, 1993). In other contexts, 'participation' is about democracy and citizenship rather than a set of methods carried out by development officials.

A related and crucial point is that, in order to see examples of progressive change, we may need to look outside conventional development and 'the project'. If development has a tendency to nullify radical ideas via its anti-politics machinery, radical and emancipatory understandings of power, participation and gender continue to thrive, albeit in different spaces. In developing on his concept of 'post-development' for example, Escobar has identified radical discourses and strategies among social movements in countries such as Ecuador, Bolivia and Venezuela that 'suggest radical possibilities towards post-liberal, post-developmentalist, and post-capitalist social forms' (Escobar, 2010). For Escobar the idea of 'post' refers to multiple 'decenterings': of the centrality of capitalism to definitions of economy, of liberalism to our definitions of society and politics, and of state power as the main framework of social organisation. This loosening is one in which new forms of the present are constantly under construction based on the idea that 'the economy is not essentially or naturally capitalist, societies are not naturally liberal, and the state is not the only way of instituting social power as we have imaged it to be' (2010: 12).

While Escobar's utopian vision may be dismissed by some as unrealistic, both big and little D/development remains alive and kicking over much of the world, both as an ideology and a set of practices. Let us finally turn to some brief examples which illustrate the continued role of anthropology in critique and action, as well as the role of anthropological methods in forming new ways of working within development.

Case 1: Using ethnography in policy settings

Ideas with anthropological roots such as participation have at times influenced development practice, as we saw in Chapter 2. As we saw also in Chapter 4, efforts have been made by donors to try to enable greater 'ownership' of development assistance as part of national policy in the form of 'sector-wide approaches'. All this has provided new opportunities for forms of anthropological engagement that go beyond project and agency settings. One recent initiative in which David Lewis played an advisory role deployed a modified form of the ethnographic method. The Swedish government's health and education 'reality check' approach in Bangladesh was a five-year initiative that drew on at least three different traditions for its inspiration – anthropological participant observation, participatory methods and the medical research 'listening study' tradition.[3]

The aim was to understand how national policy reform programmes intended to improve health and education services were being experienced by ordinary people, and then to feed these experiences back in a structured way to government and donor policy-makers who could then make changes and improvements. The approach involved putting together specially trained study teams to visit and stay with a small sample of households around the country for five days each year, getting to know them, listening to them and initiating conversations about their experiences, and observing their interactions with service providers. Some of the stories that emerged simply reinforced existing knowledge and assumptions, but in other cases received wisdom was challenged, such as on the complex causes of school drop-outs, or the wider than expected gap between the presumed rights of citizens and people's capacity to operationalize these rights.

However, policy-makers generally found it difficult to engage with primarily qualitative story-based information, which ran counter to the norms of quantitative large-scale data which many claimed should form the basis for proper 'evidence-based' policy-making. The reality check experience nevertheless suggested that 'methodological populism' – the anthropological stance of seeking a 'local' point of view (Mosse and Lewis, 2006) – offers a potentially valuable counter-narrative to the technocratic mainstream of development policy, in which people rather than numbers/measurement are returned to centre stage. It also made apparent the inadequacy of the linear-technical model of policy-making that renders power, politics and inequality invisible (see Shore et al., 2011) and shows that we cannot be naïve in the ways we think about an interface between research, evidence and policy. Placing the 'right' information in front of policy-makers is unlikely to be a major factor in changing policy ideas or implementation. Perhaps the best hope is that it can influence the general climate of ideas within which policies are devised.

In the next example we remain in Bangladesh, but with different players: this time the development organisation was not a Swedish government development agency, and nor was it particularly interested in listening closely to the poor or using ethnographic methods. Instead, it was the American multinational Chevron, which was contracted to extract natural gas from a large site at Bibiyana, in the north-east of the country. As part of the company's 'community engagement programme', a range of projects was set up, all of which played homage to the ideals of 'participation', 'partnership' and 'empowerment'. As we shall see, the old questions of access, control and effects were central to research contracted by the DFID-ESRC Joint Programme of Poverty Focussed Research and carried out by Katy and a team of researchers from Jahangirnagar University.

*Case 2: Old questions in new settings – corporate social responsibility and
extractive industries in Bangladesh*

As we saw in Chapter 3, anthropologists have increasingly been turning
their attention to the practices and discourses of extractive industries,
many of which are now touting policies of 'corporate social responsibility'
(CSR). In some cases these anthropologists have become directly involved
in activism and advocacy (see, for example, Kirsch, 2010a, 2010b;
Sawyer, 2005). Stuart Kirsch, for example, has argued that in contexts of
environmental destruction and other abuses there is no space for neutrality
(2010a, 2010b; see also 2014). Other work highlights the complex moral
orders underlying ideas of 'ethical business', not to say the anti-politics of
programmes which claim 'partnership' or 'participation' (see, for example,
Welker, 2012, 2014; Shever, 2010; Rajak, 2011). In the following example,
we see how these ideas might translate in engaged settings:

Over the period 2008–11, a team of researchers from Jahangirnagar
University Fatema Bashir, Masud Rana and Dr Zahir Ahmed, and Katy
Gardner conducted detailed, qualitative research in Bibiyana which
examined the impact of the gas field on local livelihoods and poverty.
They were particularly interested in Chevron's programme of 'community
engagement' and their claims of 'partnership', which they made in
statements on websites, in PR material, CSR reports and so on. The gas
field had been operating since 2007; in order to develop the site, 50 acres
of land, plus more for connecting roads, was forcibly acquired by the
government. The loss of land in a fragile rural economy in which large
numbers of very poor people depend on agriculture for their livelihoods
was therefore impacting the poorest, as well as the families who owned it.
There were also major changes to the environment. Due to an embankment
on the Kushiara River and the high-banked roads built for the gas field,
water no longer flowed across the land during the wet season, leading
to mechanised irrigation systems and a shift in cropping patterns. These
environmental issues, plus concerns over the safety of the installation were
crucial for local people.

While originally there was widespread protest against the gas plant by
the time of the site's inauguration in 2007 the opposition had become
fragmented; eventually some of the leaders decided to work with, rather
than oppose Chevron, who were offering to implement a package of
community development programmes. Their negotiation of improved rates
of land compensation helped, as did the prospect of work for hundreds of

local people employed as labourers on the site. By the time construction was completed in 2007, however, most of the labourers were laid off. Gas plants require highly skilled labour but not much else. As for the connection to the gas supply that local people demanded: it was not to be. Without the technology to convert the gas for domestic use on-site, the gas is pumped away from the area to be processed elsewhere.

So: no gas and no jobs. And while those who owned the land received compensation (though many complained about the bribes they had to pay to land registry officials in the process) those who depended upon it for their livelihoods – the sharecroppers, agricultural labourers and poor women and children who used it for grazing cattle, collecting straw for housing, or fishing in the wet season – were the net losers. It was these people whom Chevron claimed to be assisting in their 'community engagement' programmes, which included a range of 'alternative livelihood' projects, health clinics, education stipends and so on, contracted to the NGO, Friends in Village Development, Bangladesh to carry out. The aim of these works, a top Chevron official explained, was 'to empower people'.

As we heard in Chapter 3, anthropologists in other contexts (and this one) have shown how CSR can be understood as a means for mining companies to buy a 'social licence' to operate in places where their presence has met with opposition (see Kirsch, 2006; Zalik, 2004; Welker, 2009, 2014; Shever, 2010; Rogers, 2012). While presented as ethical business practice, such gift-giving can, to this extent, be understood as a form of bribery. One of the aims of the research was to examine whether CSR or programmes of 'community engagement' help to combat poverty. The old questions of access (to the benefits of the programmes), control (over resources, including the programmes) and effects (of the gas field, and of the programmes) were central, but with a new spin: were these programmes simply covering up dispossession, environmental damage and new forms of imperialism?

What the research showed was that, like most NGO programmes in Bangladesh (as indeed, anywhere), the projects that Chevron funded had had mixed results. Some people benefited from the trainings in goat or duck keeping, beef fattening or poultry raising, or from the credit for small business enterprises or the services offered by the clinic. Others, however, told the researchers they had not been included in the programmes. Indeed, the research suggested that, rather than shifting existing power relationships and hierarchies, the programmes played into and exacerbated them, for the benefits of the CSR initiatives were dispensed via 'village development organisations' run by members of the local elite. This meant that patron–client relationships between the wealthy and the poor remained largely unchallenged.

What about the much touted 'partnership' with the people of Bibiyana? While it was true that some 'leaders' were indeed working closely with Chevron on their programmes during the research, others – often those who had most vehemently opposed the loss of land and construction of the gas field – were excluded. To this extent, the image of 'community partnership' propagated by the company's literature was misleading, for Chevron were only 'partners' with *some* leaders and only *some* of the poor were benefiting from their programmes, often those who already had social connections with well-placed leaders who gained the benefits.

Other initiatives, which one might think would arise within a true partnership, did not exist. There were, for example, no established channels of communication between local people and company staff. Again, only those members of the local elite with close relationships with officials had access to them. Likewise, there were no formal grievance procedures or regular open meetings in the villages surrounding the site. Local people were terrified of a major accident or 'blow-out', as had happened at other gas fields in Sylhet, but safety procedures were not adequately communicated and, when planned flaring took place, only the so-called 'leaders' were warned, since it was assumed that these 'leaders' were in touch with 'the people', which they were not.

Crucially, too, Chevron Bangladesh was not transparent in its affairs. When Katy Gardner asked to see their environmental, health and social impact assessments she was told that these were internal documents, not available for public scrutiny. Likewise, there was no way of knowing the actual content of their production share contracts with the Bangladeshi government, for they were kept secret. This was all the more surprising given Chevron's claims for transparency on the global stage. The company is, for example, a signatory to the Extractive Industries Transparency Initiative.[4] Indeed, figures released recently by Revenue Watch/Transparency International make salutary reading. As they show, companies such as Chevron have reasonable track records in accountability and transparency in their 'in-house' affairs and the places where they are based (in Chevron's case, the US) but are appallingly bad at reporting anti-corruption measures and transparency initiatives in the global South.[5]

This led the researchers to question the true nature of 'partnership'. During the research they saw various philanthropic projects which chimed with fashionable ideas of sustainable development, yet no evidence of what they believed constituted true partnership with the poorest people of Bibiyana. Instead, the programmes were used, first and foremost, to boost the company's global reputation for socially responsible business practice.

What to do? The team argued that for Chevron to show true partnership it might consider the following ideas:

- *transparency*: publish the details of production share contracts and company profits, and make environmental, health and social impact assessments publicly available;
- *accountability*: ensure that grievance procedures are established, and that local people have proper redress for damages caused by company operations;
- establish proper channels of *communication* in order to convey safety information in an appropriate manner (for example, open public meetings, community liaison staff who are available to everyone, and house-to-house visits);
- establish *representative forums* so that community concerns are heard;
- support the government and civil society organisations in taking *anti-corruption measures*, and building greater transparency.

Did Chevron listen? Perhaps not, but other NGOs in Bangladesh did, with one of the largest organisations in the country telling us they had decided not to work with an energy multinational partly on the basis of our research.

If Chevron's programme shows a distinct misunderstanding of the meanings of 'empowerment' and 'partnership', is it possible for the true potential of these terms to be realised in other settings? In the next example we see how the critique of policies of GAD discussed in the last chapter, and of the 'dumbing down' of empowerment, has helped to develop more radical work which breaks away from the formulations of projects and policies which bestow 'empowerment' on beneficiaries.

Case 3: Rethinking empowerment and 'gender and development'

As we have seen, anthropologists have become increasingly critical of the use of the term 'empowerment'. Similarly, recent critiques of 'gender and development' imply that work in this field has increasingly lost its way, becoming depoliticised and turned into a series of bureaucratic exercises. Other work, however, suggests newly invigorated directions. In their introduction to a volume of essays written by participants in the Research Programme Consortium 'Pathways of Women's Empowerment' for example, Cornwall and Edwards (2010) argue that the programme's research into the different routes which women from around the world take towards empowerment cast the notion of empowerment in a radically different light. Rather than being an 'outcome' produced by development agencies and projects, women's empowerment often arises in unexpected places and may involve activities and arenas which are a long way outside the normal remit of conventional policy and projects. Indeed, rather than the quick and easy solutions and hoped-for 'outcomes' of development, processes of empowerment are exactly that: *processual*, involving choices, negotiations and narratives. Rather than seeking a linear process in which a particular policy or project would lead to a particular change (for example, electoral reform or quotas), the authors argue for policies which create deeper, structural change and a recognition that empowerment takes place over time. While educational or legislative reforms or economic initiatives remain important, most of the pathways which the programme investigates involve women strategizing and mobilising for themselves. As they put it:

> Our research reveals pathways of empowerment that lie beyond the conventional gaze of development agencies. These reveal other dimensions of empowerment in aspects of women's lives often obscured by the materialism of development: the solace of belief and the sociality of religious practice, the pleasures of leisure, and the centrality to women's lives of affective and supportive relationships with others. (Cornwall and Edwards, 2010: 2)

Within this analysis power/empowerment is not something that can simply be bestowed on others, because it is partial, contextually dependent and may be found in unexpected places, at least for Western liberal feminists. Research into Islamic belief and practices among Bangladeshi women, for example, shows how their faith brings a sense of fulfilment and self-respect, in which they are very much exerting their agency even if not directly challenging patriarchal structures. Rather, they continually negotiate between positions, and their religiosity helps bring a sense of personal empowerment (Huq, 2010).

Perhaps one of the most surprising, and radical directions of this work in a world dominated by mainstream economic understandings of 'development' is the emphasis on pleasure, leisure and affect. For example, researchers have investigated how more positive and less normative understandings of women's sexuality can enhance empowerment. While most development work has approached sexuality in negative terms, associating women's bodies with sexual abuse, violence and ill health, other approaches take pleasure and sexuality as a starting point in promoting equality and empowerment (Jolly et al., 2013). Research into the use of mobile phones and the internet in India, for example, shows that new forms of technology can be used for finding love, sex and social mobility, and can thus have empowering effects (Ganesh, 2010). Meanwhile, it is only recently that anthropologists have pointed out that gender and development involves men as much as women (Cornwall et al., 2011).

The implications of the 'Pathways' programme are far reaching: it is not simply that 'empowerment' is never the gift of feminist policy-makers or project workers, but also that sources of power are diverse, shifting, relational and – crucially – often hidden. While this may seem obvious to anthropologists accustomed to digging beneath the taken-for-granted to seek out buried meanings and sub-texts, it remains a lesson that many policy-makers and practitioners have yet to learn. Ultimately the aim of programmes such as 'Pathways' is not so much to change or implement mainstream policy as to share research and findings, creating networks for future mobilisation and activism.

This returns us to one of our central points: activist/applied/engaged anthropology is as dynamic as ever, but since we wrote *Anthropology, Development and the Post-Modern Challenge* 18 years ago, anthropologists have moved increasingly outside formalised development work and institutions in order to effect change. This does not mean that their ideas and methods are less important. Ethnography and critique are as salient as ever, as is the close attention paid to questions of power, equity and effects. Rather, the spaces in which anthropologists work have shifted. Partly this is because the inner workings of aid and 'big D' development have changed: budgetary transfers, fiscal policies and concerns over 'security' have overtaken 'the project' and the role of social advisers has shrunk. But it is also because, as we have described in these pages, the world has moved on: corporations and billionaires are as likely to instigate schemes of improvement as the conventional donors, while countries such as India and China have become 'developers'. Poverty, inequality and injustice continue unabated, but in the wake of global financial crisis new social political movements have sprung up to challenge the structures which produce them. It is into these spaces that anthropologists in the early twenty-first century must move. Indeed, as anthropologists such as David Graeber (2011) successfully use anthropological theory and ethnography to challenge the deeply engrained assumptions of classical economics, and the call for protest or anarchic anthropology grows (Maskovsky, 2013)[6] the need for anthropological critique and engagement remains as important as ever.

Conclusion: Anthropology, Development and Twenty-First-Century Challenges

Bangladesh, 24 April 2013: A tower block on the outskirts of Dhaka collapses. The building contains five garment factories and over 1,100 people are killed; over the next two weeks a further 2,438 people are evacuated, many of whom are left with debilitating injuries. Almost all are garment factory workers, the cheap labour that has enabled the country's recent high growth rates and claims of 'development success'. The 'ready-made' garments sector accounts for 80 per cent of the country's exports, providing bargain-basement clothes for many of the biggest fashion stores on the high streets and malls of Europe and the US, and employs over 4 million people, the majority of whom are young women.[1]

The disaster, which leads to intense national mourning and an international outcry over the human cost of cheap clothes in the West, is man-made in every sense. In the immediate aftermath it emerges that the day before the disaster the workers noticed cracks on the walls; the building was briefly evacuated but the next day they were ordered to continue work, even though other employers using the building, such as BRAC Bank, ordered *their* employees to stay away. Huge generators, used for the constant power cuts that characterise modern life in Bangladesh, are thought to have contributed to the collapse of the eight-storey block, which, like many factory buildings in the urban sprawl, was built on illegally seized land. The owner of the block, part of the criminal underworld associated with land grabs and illegal building, and an associate of the Awami League ruling political party, is accused of homicide for disregarding the crack and ordering workers to continue at their machines and arrested. It is not the first factory accident to claim hundreds of lives in Bangladesh. Only the year before, over a hundred workers were killed when their illegally built factory was engulfed in flames. They were unable to escape: the windows were barred and the doors locked.[2]

The international consternation over these incidents focuses not only upon the appalling conditions of factories in countries such as Bangladesh but the culpability of international firms and, indeed, consumers in the West, whose hunger for affordable fashion has a direct impact upon wages and conditions in the global South. Global interdependence is once more made explicit: international markets require cheap labour to manufacture clothes for mass consumption of fashion brands, while Bangladesh desperately needs a market for its exports.[3] Meanwhile ethical concerns – again driven partly by the market (or 'reputation') – lead buyers such as the US Disney Corporation to pull out of Bangladesh. Others announce their support of initiatives to enforce safety standards and workers' rights, which become a new focus for donors and NGOs alike: DFID, for example, pledges £4.8 million towards the National Action Plan on Fire Safety and Structural Integrity,[4] made up of around 150 companies, mainly from Europe, while Nobel prize-winning Muhammad Yunus calls for an international minimum wage.[5]

Control, access, effects

The Rana Plaza disaster tragically exemplifies all of our key themes: the lack of control that the victims had over working conditions and, indeed, the power of factory owners to break regulations and enforce inhumane working conditions; questions concerning the relationship between illegality, 'crony capitalism' (Feldman and Geisler, 2012), brute power and access to land and other resources in the Bangladeshi context, not to say the costs of access to global markets; the effects of rampant urbanisation, corruption, cost cutting and oppressive labour regimes, horribly embodied by the thousands crushed and mutilated by the collapsed tower. All can be traced to processes that we might describe as little d development, defined, as we outlined in the 'Prelude', by Gillian Hart (2001: 650) as: *'the development of capitalism as a geographically uneven, profoundly contradictory set of historical processes'*. All too might be affected by the policies and actions of big D Development: *'a post-second world war project of intervention in the "third world"'* (2001: 650): schemes that help enforce labour rights and the empowerment of factory workers, funding offered by donors that helps improve safety standards and accountability, the role of ethics and morality in contemporary business, improved infrastructure, and so on. Wolfgang Sachs was wrong when he declared in 1992 that development was in 'ruins' (1992: 1).

As we have argued throughout this book, this means that anthropology of, and in, D/development is also thriving; indeed, some of the most exciting research described in Chapter 3 addresses the issues raised by the Rana Plaza tragedy head on: the ethics and corporate social responsibility of multinationals and fair trade, the shift from field to factory and, of course, rising inequalities in contexts of economic 'development'. Rather than succumbing to what we originally glossed as 'the post-modern challenge', anthropologists have continued to grapple with big questions concerning the nature of our rapidly changing world, using their work both as 'critique' and for action, or what audit culture in higher education in the UK terms 'impact'. In many ways then, 'the post-modern challenge' has been superseded by more pressing issues. While in the early 1990s the discipline was concerned with the politics of identity and representation, troubling histories of colonial collusion and ongoing accusations of neo-imperialism, some of these problems have been overcome by new writing techniques, reflexivity and a heightened awareness of the politics surrounding the generation and dissemination of knowledge about 'the other'. More generally, global events have taken over, as they have a habit of doing. In the opening decade of the twenty-first century, two more collapsing buildings, this time in the shape of the World Trade Center in New York in 2001, led to a profound reshaping of international relations. Urgent questions concerning war, terrorism, religion, global inequalities and rights – to name just a few – meant that anthropologists had less time for solipsistic recrimination or literary deconstruction.

As we have also seen, in the last 15 years the world has been changing so fast that most anthropologists find it difficult to avoid questions of 'development', even if they do not consider themselves to be anthropologists *of* development. Indeed, as we outlined in Chapter 3, some of the most exciting work in the contemporary period focuses on global interrelationships, new economic forms, protest and the spread of neoliberal capital, as well as 'globalisation from below' (Mathews and Vega, 2012). A tiny international elite now own an ever growing share of the world's wealth,[6] making questions of poverty, inequality and access to the world's resources increasingly urgent. Meanwhile the global financial meltdown of 2008 onwards, austerity and recession, plus heightened inequalities 'at home' in Europe and the US have meant that these questions are everywhere. As we suggested in Chapters 1 and 4, the old divisions of the world into

'North and South' or 'First, Second and Third' are increasingly blurred and unhelpful. Development – using Hart's definition of the uneven spread of capital and its effects – is all around us, and impossible to ignore. Rather than simply being a colonial project 'out there', it is 'here'. When a modern consumer chooses to buy – or not to buy – a garment manufactured by a low-paid worker in Bangladesh he or she is implicated and connected with D/development through complex but inescapable chains and linkages to a basic set of issues around global poverty and justice in which we are all participants.

Anthropologists ignore these issues at their peril. While the study of 'schemes of improvement' and 'Aidnography' has been useful, it is time for a newly invigorated anthropology of development that places poverty and inequality at the centre of the enquiry. Who wins and who loses? *Why*? These are crucial questions that, as we argued in 1996, anthropologists are perfectly placed to answer. To do this, we need to continue the critique of new forms of development – corporate social responsibility, celebrity philanthropy, environmental issues, the entry of new donors and so on – but also break free from the confines of projects and schemes of improvement. As we have seen, these have a tendency to de-politicise radical ideas and might distract us from other forms of change, such as the social and political movements described in the previous chapter. At the same time, between the worlds of development projects and those of social movements lies another field of power and governance – that of 'policy worlds' – to which anthropologists are once again returning. By problematizing the concept of 'policy' as a key organising concept in the contemporary world, and making policy visible as a political and ideological construct, traditional anthropological concerns such as institutions, ideology, discourse and identity are reinvigorated (Shore and Wright, 1997).

Perhaps we also need to stop worrying about the term 'development' and simply call for an anthropology of engagement (see Eriksen, 2005) in which questions of access, control and effects in whatever parts of the world are foregrounded. To this extent, rather than being corralled with 'the anthropology of development' the twenty-first-century challenges we outline below are relevant for the whole of the discipline. What is our role in the world we work in? While most of anthropology rightly remains academic (for our primary objective is to generate knowledge of different cultures and societies – however defined – and by so doing de-centre hegemonic orthodoxies and assumptions, an enterprise which involves academic research) a second but no less important role is to engage constructively in our problematic and troubled world, work which involves both critique and action (Hale, 2006). As we have argued in both this and the earlier edition of our book, there is a role for both within the discipline. Indeed, they are mutually beneficial. Action feeds off, and is partly dependent upon academic knowledge and critique, and academic knowledge and critique is enriched by action, for as Gow (2002) has argued, the latter gives anthropology a 'moral narrative'; it prevents it from ossifying and turning inwards. If configured in terms of engagement rather than narrow questions concerning Big D development, it keeps it alive.

So, what *are* the twenty-first-century challenges for the anthropology of d/Development?

First, *to document and explain continuing and deepening inequality at all scales*. Despite economic growth among the BRICs, the economic divide between the haves and have-nots is greater than ever. The OECD reported in 2011, for example, that India's income inequality had doubled in the last 20 years,[7] while Britain had the dubious distinction among wealthy countries of having the greatest income divide, with the top 10 per cent having incomes 12 times greater than the bottom 10 per cent.[8] The focus on inequality pushes us to further critique the idea of 'development', and to study 'non development' spaces, including work in the 'North' as well as the global South.

Second, *to identify, analyse and challenge the anti-politics of development*. As we have seen, disconnected or, in Eleychar's (2002) words, 'anti-development' development continues to thrive, taking what were once radical ideas and putting them to work for neoliberal capital, be this in the guise of corporate social responsibility or World Bank programmes of entrepreneurship. Anthropologists must continue to reveal what lies behind these schemes of improvement – whether they involve 'partnership', 'self-help' or 'fair trade' – taking nothing for granted and using ethnography and theory to challenge the new orthodoxies, however sacred they may seem.

Third, *to challenge normative frameworks*, for example of sexuality, gender, race and Western secularism. For example, anthropology has much to contribute to discussions and policies surrounding 'gender and development', which foreground a Western-centric secular vision of empowerment or change which does not take into account alternatives, as evidenced by the 'Pathways of Empowerment' Programme. Likewise, the challenges to economic orthodoxy initiated by scholars such as David Graeber need to continue and intensify.

Finally, *to describe alternative ways of seeing and doing*, which aim to improve the wider wellbeing of populations but move beyond growth, development and modernisation. This might include highlighting new forms of social and political solidarity such as the Via Campesina, or the Occupy movements.

Notes

Prelude: Development, post-development and more development?

1. See: http://www.halftheskymovement.org/ (accessed 13 November 2113).
2. See: http://www.un.org/millenniumgoals/ (accessed 13 November 2013) and http://www.un.org/en/ecosoc/about/mdg.shtml (accessed 28 July 2014) for details.
3. However, in order to maintain the clarity of the text, we have tried to avoid too much unnecessary capitalization. We hope that the distinctions between these different meanings of development will be clear to the reader from the context.

1. Understanding development: Theory and practice into the twenty-first century

1. 'UK aid: is the 0.7% target still relevant?', *The Guardian* 20 March 2013 (accessed 4 April 2014).
2. Personal communication, Yuko Suda, Toyo University, Tokyo.
3. 'Gross national happiness in Bhutan: the big idea from a tiny state that could change the world', *The Guardian*, 1 December 2012 (accessed 16 April 2014).
4. Despite this changing landscape, OECD countries continue to deliver $120 billion in aid each year.

2. Applying anthropology

1. However, evolutionist ideas lived on in the modernisation theories of economic development and cultural change propagated by Rostow (1960b) and others who advocated the idea of countries passing through 'stages of growth' towards modernity.
2. *Human Organisation* has always carried articles that include applied research in the domestic settings of the US and elsewhere, as well as work concerned with international 'development' settings. In this way applied anthropology has avoided the 'us and them' distinction that tends to construct separate parallel worlds of development and non-development – and is a precursor of the more inclusive view of development we argue for in this book.
3. We are grateful to James Fairhead for drawing our attention to this.
4. There were exceptions. The founder of Mozambique's liberation movement, Dr Eduardo Mondlane, had a PhD in sociology and anthropology from Northwestern University, Illinois. Mondlane was influenced by the idea of

combining social science and political action (M. Harris, 1991: 336). Jomo Kenyatta, leader of Kenya's Mau Mau liberation movement and the country's first president, had studied anthropology under Malinowski at LSE.

5. Over time, a subtle change took place in the meaning of 'applied' work. For Malinowski, it had simply been about anthropology's relevance and application. Only much later did application come to be seen as 'secondary and derivative to theorising' (Mills, 2006: 57): 'The primary focus on academic knowledge creation allowed anthropologists to distance their practice from that of colonial administrators and curious travellers, creating a disciplinary "comfort zone" around their work' (Mills, 2006: 57). A need to protect anthropology and build its viability as a discipline explains much of the unease shown towards by academic colleagues towards their applied counterparts.

6. Angela Cheater's (1986) introduction to anthropology is a good example of a distinctive practical approach to anthropology that was developed in Zimbabwe.

7. James Fairhead usefully reminds us that these criticisms have also regularly been levelled at the non-applied anthropologist community as well (personal communication).

8. See: http://www.theasa.org.

9. In a sense this brings things full circle, because as James Fairhead (personal communication) reminds us, the very first anthropologists in the UK and the US were activist anthropologists, working for Indian rights, and the abolition of slavery in the UK and in the US.

3. The anthropology of development

1. For a summary of the debates see Olivier de Sardan (2005); Edelman and Haugerud (2005); Axelby and Crewe (2013).

2. For an account of structuralism in British social anthropology, see Kuper (1983).

3. In, for example, *Political Systems of Highland Burma* (Leach, 1954).

4   An early example of such an approach is Peter Worsley's *The Trumpet Shall Sound* (1957), an analysis of Melanesian cargo cults, which Worsley argues developed as a reaction to white colonisation during the Second World War.

5. Such as Weiner's (1976) re-evaluation of Malinowski's work on the Trobriand islanders.

6. For example, Rosaldo and Lamphere (1974), Reiter (1975), Ortner and Whitehead (1981).

7. For a wider discussion of this literature, see Kabeer (1994).

8. While WID refers to women's role in development, GAD refers to the relationship between development and socially constructed gender relations,

thus recognising historical and cultural particularities of women's (and men's) social roles and statuses.

9. For a summary of policies aimed at gender relations within development, together with a discussion of gender training, see Moser (1993).

10. As useful shorthands offering an anthropological characterisation of the big D development universe, we have mostly used Development World (Axelby and Crewe, 2013) and Aidland (Apthorpe, 2011) interchangeably throughout this book.

11. Though see Escobar's elaboration on development interventions in *Columbia: Territories of Difference* (2008).

12. After the change of government from Conservative to Labour what was the ODA became the Department for International Development in 1997.

13. For wider discussion of the CSR of Mining Corporations see Kirsch (2010a, 2010b), Shever (2010), Welker (2014), Zalik (2004), Rogers (2012).

14. See: http://pcworld.about.net/od/currentevents/Gates-Creative-Capitalism-C.htm (accessed 20 May 2013).

15. See: http://www.kiva.org/lend.

16. On humanitarian goods and their 'bio-political imaginary', see Redfield (2012).

17. Though the extent to which this is new is debateable. The East India Company, for example, was in many ways a prototype for contemporary multinationals (Robbins, 2004 – the corporation that changed the world).

18. See also the work on 'green grabbing' by James Fairhead et al. (2012) that draws on David Harvey's work to understand how the acquisition and deployment of green commodities is an important emerging process.

19. See Gow (2002), and Hale (2006) for further discussion.

4. Anthropologists in development: access, effects and control

1. Examples might be the reduction of anthropological knowledge of gender relations into training packages such as the 'triple roles framework' (Kabeer, 1994: 294–98), or the solidification into bureaucratically manageable 'indigenous knowledge systems' of complex cultural differences in ways of seeing and understanding.

2. Adapted from Lewis and McGregor (1992).

3. In 2014 Albania is the third poorest country in Europe, after Moldova and Ukraine (http://www.aneki.com/poorest_europe.html, accessed 31 March 2014).

4. Adapted from Madeley (1991: 33–38).

5. As Sen (1981) has argued, famine is not the result of objective scarcity, but a failure in people's entitlement to food, which is always mediated through social and political relationships.

6. Adapted from M. Foster (1989).

7. Adapted from Rozario (1992).
8. £1 was approximately 50 taka in 1995. It is now around Tk120 in 2014.
9. Adapted from Mair (1984: 110–13).
10. Adapted from ITDG (1992).
11. Personal communication from Proshika workers, Katy Gardner, March 1993.
12. Bilateral aid refers to situations where there is only one donor country involved. Multilateral aid involves more than one country and is implemented by multilateral agencies such as the World Bank.
13. Adapted from Gardner (1997).
14. Such criteria tend to be quantitative: for example, so many hospitals built, so many nurses employed. Measuring the success of social policies such as 'empowerment' is extremely difficult, however.

5. When good ideas turn bad: the dominant discourse bites back

1. This approach builds on Sachs's *The Development Dictionary* (1992).
2. SIDA is the Swedish International Development Cooperation Agency.
3. As we saw in Chapter 2, ethnographic practices have long been modified and applied to development practice. The World Bank's *Voices of the Poor* study attempted to build up a bottom-up picture of peoples' experiences and understandings of poverty to challenge the assumptions of outside 'experts' (Narayan et al., 1999) as did Mary B. Anderson and colleagues' (2012) work documenting people's experiences of aid in *Time to Listen: Hearing People on the Receiving End of International Aid*.
4. See: http://eiti.org/
5. See: http://www.transparency.org/policy_research/surveys_indices/promoting_revenue_transparency
6. See also: http://blogs.plos.org/neuroanthropology/2011/10/15/david-graeber-anthropologist-anarchist-financial-analyst/) (accessed 1 September 2014).

Conclusion: Anthropology, development and twenty-first-century challenges

1. See: https://www.gov.uk/government/case-studies/the-rana-plaza-disaster (accessed 19 May 2014).
2. See: http://www.nytimes.com/2012/11/26/world/asia/bangladesh-fire-kills-more-than-100-and-injures-many.html?hp (accessed 19 May 2014).
3. See for example War on Want's campaign against Bangladeshi sweat shops: http://www.waronwant.org/overseas-work/sweatshops-and-plantations/sweatshops-in-bangladesh (accessed 19 May 2014).
4   See: https://www.gov.uk/government/case-studies/the-rana-plaza-disaster (accessed 19 May 2014).
5. The Alliance for Bangladesh Worker Safety is also established and attracts 26 members, mostly from the US, raising the question as to whether having two

effectively competing safety inspection industry groups is the most efficient way forward.

6. For example, in January 2014 Oxfam released a report that showed that the 85 richest people in the world are worth more than the world's poorest 3.5 billion (http://www.washingtonpost.com/blogs/wonkblog/wp/2014/01/22/10-startling-facts-about-global-wealth-inequality/; accessed 19 May 2014).

7. As reported in *The Times of India* in December 2011: http://timesofindia.indiatimes.com/india/Indias-income-inequality-has-doubled-in-20-years/articleshow/11012855.cms (accessed 13 November 2013).

8. See: http://www.theguardian.com/society/2011/dec/05/income-inequality-growing-faster-uk (accessed 13 November 2013).

Bibliography

Acemoglu, D. and J.A. Robinson (2012) *Why Nations Fail: The Origins of Power, Prosperity and Poverty*. London: Profile Books.

Afshar, H. (ed.) (1991) *Women, Development and Survival in the Third World*. Harlow: Longman.

Ahmed, A. (1992) *Post-modernism and Islam: Predicament and Promise*. London: Routledge.

Anderson, M.B., D. Brown and I. Jean (2012) *Time to Listen: Hearing People on the Receiving End of International Aid*. Boston, MA: CDA Collaborative Learning Projects.

Apthorpe, R. (2011) 'Coda – With Alice in Aidland: a seriously satirical allegory.' In D. Mosse (ed.) *Adventures in Aidland: The Anthropology of Professional International Development*. Oxford: Berghahn, pp. 199–219.

Asad, T. (ed.) (1973) *Anthropology and the Colonial Encounter*. London: Ithaca Press.

Asad, T. (1987) 'Are there histories of people without Europe? A review article', *Society for Comparative Society and History* 29(3): 594–97.

Axelby, R. and E. Crewe (2013) *Anthropology and Development: Culture, Morality and Politics in a Globalised World*. Cambridge: Cambridge University Press.

Baba, M.L. (2005) 'Anthropological practice in business and industry.' In S. Kedia and J. Van Willigen (eds) *Applied Anthropology: Domains of Application*. Westport, CT: Praeger.

Bailey, F.G. (1958) *Caste and the Economic Frontier: A Village in Highland Orissa*. Manchester: Manchester University Press.

Banerjee, A. and Duflo, E. (2011) *Poor Economics: A Radical Rethinking of the Way to Fight Global Poverty*. Philadelphia, PA: Perseus Books.

Barber, P.G., B. Leach and W. Lem (eds) (2012) *Confronting Capital: Critique and Engagement in Anthropology*. London: Routledge.

Baré, J.-F. (1997) 'Applied anthropology in France: comments from a collective survey.' In M.L. Baba and C.E. Hill (eds) *The Global Practice of Anthropology*. Studies in Third World Societies 58. Williamsburg, VA: College of William and Mary Press.

Barnett, H.G. (1956) *Anthropology in Administration*. Evanston, IL: Row, Peterson and Co.

Barnett, T. (1977) *The Gezira Scheme: An Illusion of Development*. London: Frank Cass.

Barrios de la Chungara, D. (1983) 'Women and organisation.' In M. Davies (ed.) *Third World: Second Sex*. London: Zed Books, pp. 39–61.

Batliwala, S. (2010) 'Taking the power out of empowerment: an experiential account.' In A. Cornwall and D. Eade (eds) *Deconstructing Development Discourse: Fuzzwords and Buzzwords*. Rugby: Practical Action Publishing.

Batliwala, S. and D. Dhanraj (2007) 'Gender myths that instrumentalize women: a view from the Indian frontline'. In A. Cornwall, E. Harrison and A. Whitehead (eds) *Feminisms in Development: Contradictions, Contestations and Challenges*. London: Zed Books.

Beattie, J. (1964) *Other Cultures: Aims, Methods and Achievements in Social Anthropology*. London: Routledge and Kegan Paul.

Behrends, A., S. Reyna and G. Schlee (eds) (2011) *Crude Domination: An Anthropology of Oil (Dislocations)*. Oxford: Berghahn.

Belshaw, C. (1976) *The Sorcerer's Apprentice: An Anthropology of Public Policy*. New York: Pergamon.

Benedict, R. (1934) *Patterns of Culture*. Boston, MA: Houghton Mifflin.

Bennett, J.W. (1996) 'Applied and action anthropology: ideological and conceptual aspects', Supplement: Special issue – 'Anthropology in Public', *Current Anthropology* 37(1): S23–S53.

Benthall, J. (2010) 'Islamic humanitarianism in adversarial context.' In E. Bornstein and P. Redfield (eds) *Forces of Compassion*. Santa Fe, NM: SAR Press, pp. 99–123.

Benthall, J. and J. Bellion-Jourdan (2009) *The Charitable Crescent: Politics of Aid in the Muslim World*. London: I.B. Tauris.

Black, J.K. (1991) *Development in Theory and Practice: Bridging the Gap*. Boulder, CO: Westview Press.

Blanchard, D. (1979) 'Beyond empathy: the emergence of an action anthropology in the life and career of Sol Tax.' In R. Hinshaw (ed.) *Currents in Anthropology: Essays in Honor of Sol Tax*. New York: Mouton, pp. 419–43.

Bloch, J. and M. Bloch (1980) 'Women and the dialectics of nature in eighteenth-century French thought.' In C. MacCormack and M. Strathern (eds) *Nature, Culture and Gender*. Cambridge: Cambridge University Press, pp. 25–42.

Bloch, M. (1983) *Marxism and Anthropology: The History of a Relationship*. Oxford: Clarendon.

Bornstein, E. (2005) *The Spirit of Development: Protestant NGOs, Morality and Economics in Zimbabwe*. Stanford, CA: Stanford University Press.

Bornstein, E. (2012) *Disquieting Gifts: Humanitarianism in New Delhi*. Stanford, CA: Stanford University Press.

Bornstein, E. and P. Redfield (eds) (2010) *Forces of Compassion: Humanitarianism Between Ethics and Politics*. Santa Fe, NM: SAR Press.

Boserup, E. (1970) *Woman's Role in Economic Development*. London: Allen and Unwin.

Breman, J. (1974) *Patronage and Exploitation: Changing Agrarian Relations in South Gujarat*. Berkeley: University of California Press.

Breman, J. (1999) 'The study of industrial labour in post-colonial India – The informal sector: a concluding review', *Contributions to Indian Sociology* 33(1–2): 407–31.

Breman, J. (2004). *The Making and Unmaking of an Industrial Working Class: Sliding Down the Labour Hierarchy in Ahmedabad, India.* Amsterdam: Amsterdam University Press.

Burawoy, M. et al. (2000) *Global Ethnography: Forces, Connections, and Imaginations in a Postmodern World.* Berkeley: University of California Press.

Burghart, R. (1993) 'His lordship at the cobblers' well.' In M. Hobart (ed.) *An Anthropological Critique of Development: The Growth of Ignorance.* London: Routledge, pp. 79–100.

Burton, T. (2002) 'When corporations want to cuddle.' In G. Evans, J. Goodman and N. Lansbury (eds) *Moving Mountains: Communities Confront Mining and Globalization.* New York: Zed Books, pp. 109–25.

Campbell, J.R. (2010) 'The problem of ethics in contemporary anthropological research', *Anthropology Matters* 12(1): 1–17.

Carroll, T. (1992) *Intermediary NGOs: The Supporting Link in Grassroots Development.* West Hartford, CT: Kumarian.

Carswell, G. and G. De Neve (2014) 'T-shirts and tumblers: caste, dependency and work under neoliberalisation in south India', *Contributions to Indian Sociology* 48(1): 103–31.

Cassen, R. and Associates (1986) *Does Aid Work?* Oxford: Clarendon.

Castro, P., D. Taylor and D.W. Brokensha (eds) (2012) *Climate Change and Threatened Communities.* Rugby: Practical Action Publishing.

Cefkin, M. (2011) Comment on Jane I. Guyer 'Blueprints, judgment, and perseverance in a corporate context', *Current Anthropology* 52(Suppl. 3): S25–26.

Chambers, E. (1987) 'Applied anthropology in the post-Vietnam era: anticipations and ironies', *Annual Review of Anthropology* 16: 309–37.

Chambers, R. (1983) *Rural Development: Putting the Last First.* Harlow: Longman.

Chambers, R. (1992) *Rural Appraisal: Rapid, Relaxed and Participatory.* IDS Discussion Paper 311. Brighton: Institute of Development Studies.

Chambers, R. (1993) *Challenging the Professions: Frontiers for Rural Development.* London: Intermediate Technology Publications.

Chambers, R., A. Pacey and L.A. Thrupp (eds) (1989) *Farmer First: Farmer Innovation and Agricultural Research.* London: Intermediate Technology Publications.

Chant, S.H. (2007) *Gender, Generation and Poverty: Exploring the Feminisation of Poverty in Africa, Asia and Latin America.* Cheltenham: Edward Elgar.

Cheater, A. (1986) *Social Anthropology: An Alternative Introduction.* London: Routledge.

Clifford, J. (1988) *The Predicament of Culture: Twentieth-Century Ethnography, Literature and Art.* Cambridge, MA: Harvard University Press.

Clifford, J. and G. Marcus (eds) (1986) *Writing Culture: The Poetics and Politics of Ethnography*. Berkeley: University of California Press.

Cochrane, G. (1971) *Development Anthropology*. New York: Oxford University Press.

Cohen, A. (1969) *Custom and Politics in Urban Africa*. London: Routledge and Kegan Paul.

Comaroff, J. (1985) *Body of Power, Spirit of Resistance: The Culture and History of a South African People*. Chicago: University of Chicago Press.

Comaroff, J. and Comaroff, J.L. (2000) 'Millennium capitalism: first thoughts on a second coming', *Public Cultures* 12(2): 291–343.

Comaroff, J. and J.L. Comaroff (2012) 'Theory from the South: or, how Euro-America is evolving towards Africa', *Anthropological Forum* 22(2): 113–31.

Cooke, B. and U. Kothari (eds) (2001) *Participation: The New Tyranny?* London: Zed Books.

Corbridge, S. and A. Shah (2013) 'Introduction: the underbelly of the Indian boom', *Economy and Society* 42(3): 335–47.

Cornwall, A. (2003) 'Whose voices? Whose choices? Reflections on gender and participatory development', *World Development* 31(8): 1325–42.

Cornwall, A. (2010) 'Introductory overview: buzzword and fuzzwords: deconstructing development discourse'. In A. Cornwall and D. Eade (eds) *Deconstructing Development Discourse: Fuzzwords and Buzzwords*. Rugby: Practical Action Publishing.

Cornwall, A. and D. Eade (eds) (2010) *Deconstructing Development Discourse: Fuzzwords and Buzzwords*. Rugby: Practical Action Publishing.

Cornwall, A. and J. Edwards (2010) 'Introduction: negotiating empowerment', *IDS Bulletin* 41(2).

Cornwall, A. and N. Lindisfarne (eds) (1994) *Dislocating Masculinity: Comparative Ethnographies*. London: Routledge.

Cornwall, A. and G. Pratt (2011) 'The use and abuse of participatory rural appraisal: reflections from practice', *Agriculture and Human Values* 28(2): 263–72.

Cornwall, A. and I. Scoones (eds) (2011) *Revolutionizing Development: Reflections on the Work of Robert Chambers*. London: Earthscan.

Cornwall, A., E. Harrison and A. Whitehead (eds) (2007) *Feminisms in Development: Contradictions, Contestations and Challenges*. London: Zed Books.

Cornwall, A., J. Edstrom and A. Greig (2011) *Men and Development: Politicising Masculinities*. London: Zed Books.

Coumans, C. (2011) 'Occupying spaces created by conflict: anthropologists, development NGOs, responsible investment, and mining', *Current Anthropology* 52(Suppl. 3): S29–44.

Cowan, J.K., M.-B. Dembour and R.A. Wilson (eds) (2001) *Culture and Rights: Anthropological Perspectives*. Cambridge: Cambridge University Press.

Cowan, M. and R. Shenton (1995) 'The invention of development'. In J. Crush (ed.) *Power of Development*. London: Routledge.

Craig, D. and D. Porter (2006) *Development beyond Neoliberalism? Governance, Poverty Reduction and Political Economy*. London: Routledge.

Crewe, E. and E. Harrison (1998) *Whose Development? An Ethnography of Aid*. London: Zed Books.

Cross, J. (2011) 'Detachment as corporate ethic: materialising CSR in the diamond supply chain', *Focaal – Journal of Global and Historical Anthropology* 60: 34–46.

Cross, J. and A. Street (2009) 'Anthropology at the bottom of the pyramid', *Anthropology Today* 25(4): 4–9

Darwin, C. (1956 [first published 1859]) *The Origin of Species*. London: Dent.

D'Espallier, B., I. Guérin and R. Mersland (2011) 'Women and repayment in microfinance: a global analysis', *World Development* 39(5): 758–72.

de Haan, A. (2009) *How the Aid Industry Works: An Introduction to International Development*. Sterling, VA: Kumarian.

De Neve, G. (2009) 'Power, inequality and corporate social responsibility: the politics of ethical compliance in the South Indian garment industry', *Economic and Political Weekly* 44(22): 63–71.

De Neve, G., P. Luetchford and J. Pratt (2008a) 'Introduction: revealing the hidden hands in global market exchange.' In G. De Neve, P. Luetchford and J. Pratt and D. Wood (eds) *Hidden Hands in the Market: Ethnographies of Fair Trade, Ethical Consumption and Corporate Responsibility*. London: JAI Press, pp. 1–30.

De Neve, G., P. Luetchford, J. Pratt and D. Wood (eds) (2008b) *Hidden Hands in the Market: Ethnographies of Fair Trade, Ethical Consumption and Corporate Responsibility*. London: JAI Press.

de Soto, H. (2000) *The Mystery of Capital: Why Capitalism Triumphs in the West and Fails Everywhere Else*. New York: Basic Books.

De Wet, C. (1991) 'Recent deliberations on the state and future of resettlement anthropology', *Human Organisation* 50(1): 104–09.

Dey, J. (1981) 'Gambian women: unequal partners in rice development projects?' In N. Nelson (ed.) *African Women in the Development Process*. London: Frank Cass, pp. 109–22.

Dolan, C. (2007) 'Market affections: moral encounters with Kenyan Fairtrade flowers', *Ethnos* 72(2): 239–61.

Dolan, C. (2008) 'Arbitrating risk through moral values: the case of Kenyan flowers.' In G. De Neve, P. Luetchford, J. Pratt and D. Wood (eds) *Hidden Hands in the Market: Ethnographies of Fair Trade, Ethical Consumption and Corporate Responsibility*. London: JAI Press, pp. 271–97.

Dolan, C. and M. Johnstone-Louis (2011) 'Re-siting corporate social responsibility: the making of South Africa's Avon entrepreneurs', *Focaal – Journal of Global and Historical Anthropology* 60: 21–33.

Dollar, D. and A. Kraay (2002) *Growth Is Good for the Poor*. Development Research Group report. Washington, DC: World Bank.

Dos Santos, T. (1973) 'The crisis of development theory and the problem of dependence in Latin America.' In H. Bernstein (ed.) *Underdevelopment and Development*. Harmondsworth: Penguin, pp. 57–80.

Durkheim, E. (1947 [first published 1893]) *The Division of Labour in Society*. New York: Free Press.

Eade, D. (2010) 'Preface'. In A. Cornwall and D. Eade (eds) *Deconstructing Development Discourse: Fuzzwords and Buzzwords*. Rugby: Practical Action Publishing.

Eades, J. (ed.) (1987) *Migrants, Workers and the Social Order*, ASA Monograph 26. London: Tavistock.

Easterly, W. (2006) *The White Man's Burden: Why the West's Efforts to Aid the Rest Have Done So Much Ill and So Little Good*. Harmondsworth: Penguin.

Edelman, M. (2013) 'Development.' In J. Carrier and D. Gewertz (eds) *Handbook of Sociocultural Anthropology*. London: Bloomsbury Academic.

Edelman, M. and A. Haugerud (2005) *The Anthropology of Development and Globalisation: From Classical Political Economy to Contemporary Neo-Liberalism*. Malden, MA: Blackwell.

Elyachar, J. (2002) 'Empowerment money: the World Bank, non-governmental organizations, and the value of culture in Egypt', *Public Culture* 14(3): 493–513.

Elyachar, J. (2005) *Markets of Dispossession: NGOs, Economic Development, and the State in Cairo*. Durham, NC: Duke University Press.

Elyachar, J. (2012) 'Before (and after) neoliberalism: tacit knowledge secrets of the trade and the public sector in Egypt', *Cultural Anthropology* 27(1): 76–96.

Engels, F. (1972 [first published 1884]) *The Origin of the Family, Private Property and the State*. New York: International Publishers.

Epstein, A. (1958) *Politics in an Urban African Community*. Manchester: Manchester University Press.

Epstein, T.S. (1962) *Economic Development and Social Change in South India*. Manchester: Manchester University Press.

Epstein, T.S. (1973) *South India: Yesterday, Today, and Tomorrow*. London: Macmillan.

Epstein, T.S. and A. Ahmed (1984) 'Development anthropology in project implementation'. In W.L. Partridge (ed.) *Training Manual in Development Anthropology*. Washington, DC: American Anthropological Association, pp. 31–41.

Eriksen, T.H. (2005) *Engaging Anthropology: The Case for a Public Presence*. Oxford: Berg.

Escobar, A. (1988) 'Power and visibility: development and the intervention and management of the Third World', *Cultural Anthropology* 3(4): 428–43.

Escobar, A. (1991) 'Anthropology and the development encounter: the making and marketing of development anthropology', *American Ethnologist* 18(4): 658–81.

Escobar, A. (1992) 'Culture, practice and politics: anthropology and the study of social movements', *Critique of Anthropology* 12(4): 395–432.

Escobar, A. (1995) *Encountering Development: The Making and Unmaking of the Third World*. Princeton, NJ: Princeton University Press.

Escobar, A. (2008) *Territories of Difference: Place, Movements, Life, Redes*. Durham, NC: Duke University Press.

Escobar, A. (2010) 'Latin America at a crossroads: alternative modernizations, post-liberalism, or post-development?', *Cultural Studies* 24(1): 1–65.

Escobar, A. and C. Ciobanu (2012) 'Latin America in a post-development era: an interview with Arturo Escobar', Open Democracy, 6 November; http://www.opendemocracy.net/openeconomy/arturo-escobar-claudia-ciobanu/latin-america-in-post-development-era-interview-with-artu (accessed 31 May 2014).

Esteva, G. (1993) 'Development'. In W. Sachs (ed.) *The Development Dictionary: A Guide to Knowledge as Power*. London: Zed Books, pp. 6–26.

Evans-Pritchard, E.E. (1940) *The Nuer: A Description of the Modes of Livelihood and the Political Institutions of a Nilotic People*. Oxford: Clarendon.

Evans-Pritchard, E.E. (1946) 'Applied anthropology', *Africa* 16(1): 92–981.

Eyben, R. (2004) 'Battles over booklets: gender myths in the British aid programme', *IDS Bulletin* 35(4): 73–81.

Eyben, R. (2011) 'The sociality of international aid and policy convergence.' In D. Mosse (ed.) *Adventures in Aidland: The Anthropology of Professional International Development*. Oxford: Berghahn, pp. 139–61.

Eyben, R. (2013) 'Uncovering the politics of "evidence" and "results": a framing paper for development practitioners'. Unpublished paper, Institute of Development Studies, University of Sussex. www.bigpushforward.net

Fairhead, J., M. Leach and I. Scoones (2012) 'Green grabbing: a new appropriation of nature?' *Journal of Peasant Studies* 39(2): 237–61.

Farmer, B.H. (ed.) (1977) *Green Revolution? Technology and Change in Rice Growing Areas of Tamil Nadu and Sri Lanka*. London: Macmillan.

Fassin, D. (2012) *Humanitarian Reason: A Moral History of the Present*. Berkeley: University of California Press.

Fechter, A.M. and H. Hindman (eds) (2011) *Inside the Everyday Lives of Aid Workers: The Challenges and Futures of Aidland*. Sterling, VA: Kumarian.

Feldman, S. and C. Geisler (2012) 'Land expropriation and displacement in Bangladesh', *Journal of Peasant Studies* 39(3–4): 971–93.

Ferguson, J. (1990) *The Anti-Politics Machine: 'Development', Depoliticisation, and Bureaucratic Power in Lesotho*. Cambridge: Cambridge University Press.

Ferguson, J. (1997) 'Anthropology and its evil twin? Development and the constitution of a discipline.' In F. Cooper and R. Packard (eds) *International Development and the Social Sciences: Essays on the History and Politics of Knowledge*. Berkeley: University of California Press, pp. 150–76.

Ferguson, J. (2005) 'Seeing it like an oil company: space, security and global capital in neo-liberal Africa', *American Anthropologist* 107(3): 377–82.

Ferguson, J. (2009) 'The uses of neoliberalism', *Antipode* 41(S1): 166–84.

Firth, R. (1981) 'Engagement and detachment: reflections on applying social anthropology to social affairs', *Human Organisation* 40(3): 193–201.

Folbre, N. (1986) 'Hearts and spades: paradigms of household economics', *World Development* 14(2): 245–55.

Foster, G. (1962) *Traditional Cultures and the Impact of Technological Change*. Evanston, IL: Harper and Row.

Foster, M. (1989) 'Environmental upgrading and intra-urban migration in Calcutta'. Unpublished PhD thesis, University of Nottingham.

Foucault, M. (1970) *The Order of Things: An Archaeology of the Human Sciences*, trans. A. Sheridan-Smith. New York: Random House.

Frank, A.G. (1967) *Capitalism and Underdevelopment in Latin America*. London: Monthly Review.

Freire, P. (1968) *The Pedagogy of the Oppressed*. New York: Seabury.

Friedmann, J. (1992) *Empowerment: The Politics of Alternative Development*. Oxford: Blackwell.

Fuentes-Nieva, R. and N. Galasso (2014) *Working for the Few: Political Capture and Economic Inequality*. Oxford: Oxfam GB.

Fukuyama, F. (1992) *The End of History and the Last Man*. New York: Avon Books.

Ganesh, I.M. (2010) *'Mobile Love Videos Make Me Feel Healthy': Rethinking ICTs for Development*. IDS Working Paper 352. Brighton: Institute of Development Studies.

Garber, B. and P. Jenden (1993) 'Anthropologists or anthropology? The Band Aid perspective'. In J. Pottier (ed.) *Practising Developing: Social Science Perspectives*. London: Routledge, pp. 50–71.

Gardner, K. (1997) 'Mixed messages: contested development and the Plantation Rehabilitation Project'. In R. Grillo and R.S. Stirrat (eds) *Discourses of Development*. Oxford: Berg.

Gardner, K. (2012) *Discordant Development: Global Capitalism and the Struggle for Connection in Bangladesh*. London: Pluto Press.

Gardner, K. and D. Lewis (1996) *Anthropology, Development and the Post-modern Challenge*. London: Pluto.

Gatter, P. (1993) 'Anthropology in farming systems research: a participant observer in Zambia'. In J. Pottier (ed.) *Practising Development: Social Science Perspectives*. London: Routledge, pp. 153–87.

Gaventa, J. (1999) 'Crossing the great divide: building links and learning between NGOs and community-based organizations in North and South'. In D. Lewis (ed.) *International Perspectives on Voluntary Action: Reshaping the Third Sector*. London: Earthscan.

Geertz, C. (1963) *Agricultural Involution: The Processes of Change in Indonesia*. Berkeley: University of California Press.

Giddens, A. (1971) *Capitalism and Modern Social Theory: An Analysis of the Writings of Marx, Durkheim, and Weber*. Cambridge: Cambridge University Press.

Goetz, A.M. and R. Sen Gupta (1996) 'Who takes the credit? Gender, power, and control over loan use in rural credit programs in Bangladesh', *World Development* 24(1): 45–63.

Goodenough, W.H. (1963) *Cooperation in Change: An Anthropological Approach to Community Development*. New York: Russell Sage Foundation.

Gould, J. (2004) 'Introducing aidnography'. In J. Gould and H.S. Marcusen (eds) *Ethnographies of Aid: Exploring Development Texts and Encounters*. Roskilde, Denmark: Graduate School of International Development Studies, Roskilde University.

Gow, D. (2002) 'Anthropology and development: evil twin or moral narrative?', *Human Organisation* 61(4): 299–313.

Graeber, D. (2011) *Debt: the First Five Thousand Years*. New York: Melville House.

Graeber, D. (2013) *The Democracy Project: A History, a Crisis, a Movement*. New York: Spiegel and Grau.

Green, M. (2003) 'Globalising development in Tanzania: policy franchising through participatory project management', *Critique of Anthropology* 23(2): 123–43.

Griffith, D., S. Liu, M. Paolisso and A. Stuesse (2013) 'Enduring whims and public anthropology', *American Anthropologist* 115(1): 125–31.

Grillo, R. (1985) 'Applied anthropology in the 1980s: retrospect and prospect'. In R. Grillo and A. Rew (eds) *Social Anthropology and Development Policy*, ASA Monographs 23. London: Tavistock, pp. 1–36.

Grillo, R. and R.L. Stirrat (eds) (1997) *Discourses of Development*. Oxford: Berg.

Grimshaw, A. and K. Hart (1993) *Anthropology and the Crisis of the Intellectuals*. Prickly Pear Pamphlet no. 1. Cambridge: Prickly Pear Press.

Guérin, I. (2013) 'Bonded labour, agrarian changes and capitalism: emerging patterns in South India', *Journal of Agrarian Change* 13(3): 405–23.

Hale, C.R. (2006) 'Activist research v. cultural critique: indigenous land rights and the contradictions of politically engaged anthropology', *Cultural Anthropology* 21(1): 96–120.

Hancock, G. (1989) *Lords of Poverty*. London: Macmillan.

Hann, C. and K. Hart (2011) *Economic Anthropology: History, Ethnography, Critique*. Cambridge: Polity.

Hannerz, U. (1980) *Exploring the City: Enquiries towards Urban Anthropology*. New York: Columbia University Press.

Harper, I. (2011) 'World health and Nepal: producing internationals, healthy citizenship and the cosmopolitan.' In D. Mosse (ed.) *Adventures in Aidland: The Anthropology of Professional International Development*. Oxford: Berghahn, pp. 123–39.

Harris, M. (1991) *Cultural Anthropology*, 3rd edn. New York: HarperCollins.

Harris, O. (1984) 'Households as natural units'. In K. Young, C. Wolkowitz and R. McCullagh (eds) *Of Marriage and the Market: Women's Subordination Internationally and its Lessons*, 2nd edn. London: Routledge and Kegan Paul, pp. 136–57.

Harriss, J. (1977) 'Implications of changes in agriculture for social relationships at the village level: the case of Randam'. In B.H. Farmer (ed.) *Green Revolution? Technology and Change in Rice Growing Areas of Tamil Nadu and Sri Lanka.* London: Macmillan, pp. 225–45.

Hart, G. (2001) 'Development critiques in the 1990s: *culs de sac* and promising paths', *Progress in Human Geography* 25(4): 649–58.

Harvey, D. (2005) *The New Imperialism.* Oxford: Oxford University Press.

Harvey, D. (2007) *A Brief History of Neoliberalism.* Oxford: Oxford University Press.

Hashemi, S.M. (1989) 'NGOs in Bangladesh: alternative development or alternative rhetoric?' Dhaka, mimeo. Institute of Development Policy and Management, University of Manchester.

Haviland, W.A. (1975) *Cultural Anthropology.* New York: Holt, Rinehart and Winston.

Hayter, T. (1971) *Aid as Imperialism.* Harmondsworth: Penguin.

Hill, P. (1986) *Development Economics on Trial: The Anthropological Case for a Prosecution.* Cambridge: Cambridge University Press.

Ho, K. (2009) *Liquidated: An Ethnography of Wall Street.* Durham, NC: Duke University Press.

Hobart, M. (ed.) (1993) *An Anthropological Critique of Development: The Growth of Ignorance.* London: Routledge.

Hoben, A. (1982) 'Anthropologists and development', *Annual Review of Anthropology* 11: 349–75.

Hopgood, S. (2006) *Keepers of the Flame: Understanding Amnesty International.* Ithaca, NY: Cornell University Press.

Howard, M.C. (1993) *Contemporary Cultural Anthropology*, 4th edn. New York: HarperCollins.

Howell, J. (2006) 'The global war on terror, development and civil society'. *Journal of International Development* 18(1): 121–35.

Huq, S. (2010) 'Negotiating Islam: conservatism, splintered authority and empowerment in urban Bangladesh', *IDS Bulletin* 41: 97–105.

Jenkins, R. (2005) 'Globalisation, CSR and poverty', *International Affairs* 81(3): 525–40.

ITDG (Intermediate Technology and Development Group) (1992) *Working with Women in Kenya.* London: ITDG.

James, W. (1973) 'The anthropologist as reluctant imperialist'. In T. Asad (ed.) *Anthropology and the Colonial Encounter.* London: Ithaca Press, pp. 41–69.

James, W. (1999) 'Empowering ambiguities'. In A. Cheater (ed.) *The Anthropology of Power*, ASA Monographs 36. London: Routledge.

Johannsen, A.M. (1992) 'Applied anthropology and post-modernist ethnography', *Human Organisation* 51(1): 71–81.

Jolly, S., A. Cornwall and K. Hawkins (2013) *Women, Sexuality and the Political Power of Sexuality.* London: Zed Books.

Jordanova, L.J. (1980) 'Natural facts: a historical perspective on science and sexuality'. In C. MacCormack and M. Strathern (eds) *Nature, Culture and Gender*. Cambridge: Cambridge University Press, pp. 42–70.

Kabeer, N. (1994) *Reversed Realities: Gender Hierarchies in Development Thought*. London: Verso.

Kabeer, N. (2002) *The Power to Choose: Bangladeshi Women and Labor Market Decisions in London and Dhaka*. London: Verso.

Kapelus, P. (2002) 'Mining, CSR and "the community": the case of Rio Tinto, Richards Bay Minerals and the Mbonambi', *Journal of Business Ethics* 39: 275–96.

Karim, L. (2011) *Microfinance and Its Discontents: Women in Debt in Bangladesh*. Minneapolis: University of Minnesota Press.

Karlsson, B.G. (2013) 'Writing development', *Anthropology Today* 29(2): 4–7.

Kedia, S. and J. van Willigen (2005) *Applied Anthropology: Domains of Application*. New York: Praeger.

Kerr, C., J.T. Dunlop, F.H. Harbison and C.A. Meyers (1973) *Industrialisation and Industrial Man: The Problems of Labour and Management in Industrial Growth*. Harmondsworth: Penguin.

Khan, M., D.J. Lewis, A.A. Sabri and M. Shahabuddin (1993) 'Proshika's livestock and social forestry programmes'. In J. Farrington and D.J. Lewis (eds) *Non-government Organisations and the State in Asia: Rethinking Roles in Agricultural Development*. London: Routledge, pp. 59–66.

Kirsch, S. (2002) 'Rumour and other narratives of political violence in West Papua', *Critique of Anthropology* 22: 53–79.

Kirsch, S. (2006) *Reverse Anthropology: Indigenous Analysis of Social and Environmental Relations in New Guinea*. Stanford, CA: Stanford University Press.

Kirsch, S. (2010a) 'Sustainability and the BP oil spill', Guest Editorial, *Dialectical Anthropology* 34(3): 295–300.

Kirsch, S. (2010b) 'Sustainable mining', *Dialectical Anthropology* 34(3): 87–93.

Kirsch, S. (2014) *Mining Capitalism: The Relationship between Corporations and Their Critics*. Berkeley: University of California Press.

Korten, D. (1990) *Getting to the 21st Century: Voluntary Action and the Global Agenda*. West Hartford, CT: Kumarian.

Kothari, U. (2005) 'A radical history of development studies: individuals, institutions and ideologies'. In U. Kothari (ed.) *A Radical History of Development Studies*. London: Zed Books.

Kramsjo, B. and G. Wood (1992) *Breaking the Chains: Collective Action for Social Justice among the Rural Poor in Bangladesh*. London: Intermediate Technology.

Kuper, A. (1983) *Anthropology and Anthropologists: The Modern British School*. London: Routledge and Kegan Paul.

Lan, D. (1985) *Guns and Rain: Guerrillas and Spirit Mediums in Zimbabwe*. London: James Currey.

Larrain, J. (1989) *Theories of Development: Capitalism, Colonialism and Dependency*. Cambridge: Polity.

Leacock, E. (1972) 'Introduction'. In F. Engels, *The Origin of the Family, Private Property and the State*. New York: International Publishers.

Leach, E. (1954) *The Political Systems of Highland Burma: A Study of Kachin Social Organisation*. London: G. Bell and Sons.

Leal, P.A. (2010) 'Participation: the ascendancy of a buzzword in the neo-liberal era.' In A. Cornwall and J. Eade (eds) *Deconstructing Development Discourse: Fuzzwords and Buzzwords*. Rugby: Practical Action Publishing, pp. 89–100.

Luetchford, P. (2008) *Fair Trade and a Global Commodity: Coffee in Costa Rica*. London: Pluto Press.

Lewis, D. (1993) 'Going it alone: female-headed households, rights and resources in rural Bangladesh', *European Journal of Development Research* 5(2): 23–42.

Lewis, D. (2005) 'Individuals, organisations and public action: trajectories of the "non-governmental" in development studies'. In U. Kothari (ed.) *A Radical History of Development Studies*. London: Zed Books.

Lewis, D. (2011a) *Bangladesh: Politics, Economy and Civil Society*. Cambridge: Cambridge University Press.

Lewis, D. (2011b) 'Tidy concepts, messy lives: defining tensions in the domestic and overseas careers of UK non-governmental professionals'. In D. Mosse (ed.) *Adventures In Aidland: The Anthropology of Professionals in International Development*. Oxford: Berghahn, pp. 177–98.

Lewis, D. (2014a) 'Contesting parallel worlds: time to abandon the distinction between the "international" and "domestic" contexts of third sector scholarship?' *Voluntas*, online first.

Lewis, D (2014b) *Non-Governmental Organisations, Management and Development*, London: Routledge.

Lewis, D. and J.A. McGregor (1992) *Change and Impoverishment in Albania: A Report for Oxfam*. Centre for Development Studies Report Series no. 1, University of Bath.

Lewis, D., G. Wood and R. Gregory (1993) 'Indigenising extension: farmers, fishseed traders and poverty-focused aquaculture in Bangladesh', *Development Policy Review* 11: 185–94.

Lewis, O. (1961) *The Children of Sanchez: An Autobiography of a Mexican Family*. New York: Random House.

Li, T.M. (2007) *The Will to Improve: Governmentality, Development and the Practice of Politics*. Durham, NC: Duke University Press.

Lloyd, P. (1979) *Slums of Hope? Shanty Towns of the Third World*. Harmondsworth: Penguin.

Long, N. (1977) *An Introduction to the Sociology of Developing Societies*. London: Tavistock.

Long, N. and A. Long (1992) *Battlefields of Knowledge: The Interlocking of Theory and Practice in Social Research and Development*. London: Routledge.

Long, N. and M. Villareal (1994) 'The interweaving of knowledge and power in development interfaces'. In I. Scoones and J. Thompson (eds) *Beyond Farmer First*. London: Intermediate Technology.

McGrew, A. (1992) 'The Third World in the New Global Order'. In T. Allen and A. Thomas (eds) *Poverty and Development in the 1990s*. Oxford: Oxford University Press, pp. 256–72.

Madeley, J. (1991) *When Aid is No Help: How Projects Fail, and How They Could Succeed*. London: Intermediate Technology.

Mair, L. (1969) *Anthropology and Social Change*. LSE Monographs on Social Anthropology 38. London: Athlone Press.

Mair, L. (1984) *Anthropology and Development*. London: Macmillan.

Malinoswki, B. (1922) *Argonauts of the Western Pacific*. London: Routledge and Kegan Paul.

Malinowski, B. (1929) 'Practical anthropology', *Africa* 2(1): 28–38.

Malinowski, B. (1961 [1945]) *The Dynamics of Culture Change: An Inquiry into Race Relations in Africa*. New Haven, CT: Yale University Press.

Mamdani, M. (1972) *The Myth of Population Control: Family, Caste and Class in an Indian Village*. New York: Monthly Review.

Mangin, W. (1967) 'Latin American squatter settlements: a problem and a solution', *Latin American Research Review* 2: 65–98.

Marcus, G. and M. Fischer (1986) *Anthropology as Cultural Critique: An Experimental Moment in the Human Sciences*. Chicago: University of Chicago Press.

Maskovsky, J. (2013) 'Protest anthropology in a moment of global unrest', *American Anthropologist* 115(1): 126–29.

Mathews, G. and C. Izquierdo (eds) (2009) *Pursuits of Happiness: Well-being in Anthropological Perspective*. Oxford: Berghahn.

Mathews, G., G. Lins Ribeiro and C. Alba Vega (eds) (2012) *Globalization from Below: The World's Other Economy*. London: Routledge.

Mathur, H.M. (1989) *Anthropology and Development in Traditional Societies*. New Delhi: Vikas.

Mauss, M. (1990 [1954]) *The Gift: The Form and Reason for Exchange in Archaic Societies*. New York: Norton.

Mawdsley, E. (2012) *From Recipients to Donors: Emerging Powers and the Changing Development Landscape*. London: Zed Books.

Mead, M. (1977) *Letters from the Field: 1925–75*. New York: Harper and Row.

Midgley, J. (1995) *Social Development: The Developmental Perspective in Social Welfare*. London: Sage.

Mills, D. (2006) 'Dinner at Claridges? Anthropology and the "captains of industry", 1947–1955'. In S. Pink (ed.) *Applications of Anthropology: Professional Anthropology in the Twenty-first Century*. Oxford: Berghahn, pp. 55–71.

Mintz, S. (1985) *Sweetness and Power: The Place of Sugar in Modern History*. New York: Viking.

Mitchell, C. (1956) *The Kalela Dance*. Rhodes-Livingstone Papers 27. Manchester: Manchester University Press.

Mohanty, C. (1988) 'Under Western eyes: feminist scholarship and colonial discourses', *Feminist Review* 30: 61–88.

Montgomery, E. and J.W. Bennett (1979) 'Anthropological studies of food and nutrition: the 1940s and the 1970s', in W. Goldschmidt (ed.) *The Uses of Anthropology*. Washington, DC: American Anthropological Association, pp. 125–34.

Moore, H. (1988) *Feminism and Anthropology*. Cambridge: Polity.

Moser, C. (1989) 'Gender planning in the Third World: meeting practical and strategic gender needs', *World Development* 17(11): 1799–825.

Moser, C. (1993) *Gender Planning and Development: Theory, Practice and Training*. London: Routledge.

Mosley, P. (1987) *Overseas Aid: Its Defence and Reform*. Brighton: Wheatsheaf.

Mosse, D. (2001) '"People's knowledge", participation and patronage: operations and representations in rural development'. In B. Cook and U. Kothari (eds) *Participation: The New Tyranny*. London: Zed Books.

Mosse, D. (2005) *Cultivating Development: An Ethnography of Aid Policy and Practice*. London: Pluto Press.

Mosse, D. (ed.) (2011a) *Adventures in Aidland: The Anthropology of Professional International Development*. Oxford: Berghahn.

Mosse, D. (2011b) 'Politics and ethics: ethnographies of expert knowledge and professional identities'. In C. Shore, S. Wright and D. Però (eds) *Policy Worlds: Anthropology and the Analysis of Contemporary Power*. Oxford: Berghahn.

Mosse, D. and D. Lewis (2006) 'Theoretical approaches to brokerage and translation in development'. In D. Lewis and D. Mosse (eds) *Development Brokers and Translators: An Ethnography of Aid and Agencies*. Bloomington, CT: Kumarian.

Moyo, D. (2009) *Dead Aid: Why Aid Is Not Working and How There is Another Way for Africa*. New York: Farrar, Straus and Giroux.

Mukhopadhyay, M. (2004) 'Mainstreaming gender or "streaming" gender away: feminists marooned in the development business', *IDS Bulletin* 35: 95–103.

Münster, D. and U. Münster (2012) 'Consuming the forest in an environment of crisis: nature tourism, forest conservation and neoliberal agriculture in South India', *Development and Change* 43(1): 205–27.

Münster, D. and C. Strümpell (2014) 'The anthropology of neoliberal India: an introduction', *Contributions to Indian Sociology* 48(1): 1–16.

Murray, C. (1981) *Families Divided: The Impact of Migrant Labour in Lesotho*. Cambridge: Cambridge University Press.

Narayan, D. with R. Patel, K. Schafft, A. Rademacher and S. Koch-Schulte (1999) *Can Anyone Hear Us? Voices From 47 Countries*. Washington, DC: World Bank.

Nash, J. (1979) *We Eat the Mines and the Mines Eat Us: Dependency and Exploitation in Bolivian Tin Mines*. New York: Columbia University Press.

Ng, C. (1991) 'Malay women and rice production in west Malaysia'. In H. Afshar (ed.) *Women, Development and Survival in the Third World.* Harlow: Longman, pp. 188–210.

Nicholson, L. (ed.) (1990) *Feminism/Post-Modernism.* London: Routledge.

Nuremowla, S. (2011) *Resistance, Rootedness and Mining Protest in Phulbari.* Unpublished thesis, University of Sussex.

OECD (2011) *Divided We Stand: Why Inequality Keeps Rising.* Paris: OECD.

Olivier de Sardan, J.P. (2005) *Anthropology and Development: Understanding Contemporary Social Change.* London: Zed Books.

Opler, M. with E. Spicer and K. Luomala (1969 [1946]) *Impounded People.* Ticson: University of Arizona Press.

Ortner, S. and H. Whitehead (eds) (1981) *Sexual Meanings: The Cultural Construction of Gender and Sexuality.* Cambridge: Cambridge University Press.

Ostergaard, L. (ed.) (1992) *Gender and Development: A Practical Guide.* London: Routledge.

Oxford Dictionary of Current English, edited by R.E. Allen (1988) Oxford: Oxford University Press.

Padel, F. and S. Das (2011) *Out of this Earth: East India Adivasis and the Aluminium Cartel.* New Delhi: Orient Black Swan.

Paiement, J.J. (2007) 'Anthropology and development', *NAPA Bulletin* 27: 196–223.

Parry, J. (1986) 'The Gift, the Indian gift and "the Indian gift"', *Man* 21(3): 453–73.

Parsons, T. (1949) 'The social structure of the family'. In R. Anshen (ed.) *The Family: Its Function and Destiny.* New York: Harper and Row, pp. 173–201.

Pearse, A. (1980) *Seeds of Plenty, Seeds of Want: Social and Economic Implications of the Green Revolution.* Oxford: Clarendon.

Perlman, J. (1976) *The Myth of Marginality: Urban Poverty and Politics in Rio de Janeiro.* Berkeley: University of California Press.

Piketty, T. (2014) *Capital in the Twenty-first Century*, trans. A. Goldhammer. Cambridge, MA: Harvard University Press.

Pitt, M. and S.R. Khandker (1998) 'The impact of group-based credit programs on poor households in Bangladesh: does the gender of participants matter?' *Journal of Political Economy* 106(5): 958–96.

Polgar, S. (1979) 'Applied, action, radical, and committed anthropology'. In R. Hinshaw (ed.) *Currents in Anthropology: Essays in Honor of Sol Tax.* New York: Mouton, pp. 409–18.

Pottier, J. (ed.) (1993) *Practising Development: Social Science Perspectives.* London: Routledge.

Power, M. (1997) *The Audit Society: Rituals of Verification.* Oxford: Oxford University Press.

Prahalad, C.K. (2004) *The Fortune at the Bottom of the Pyramid: Eradicating Poverty through Profits.* Upper Saddle River, NJ: Prentice Hall.

Price, D. (2011) *Weaponizing Anthropology: Social Science in Service of the Militarized State.* Oakland CA: AK Press/Counter Punch Books.

Rabinow, P. (1986) 'Representations are social facts: modernity and post-modernity in anthropology'. In J. Clifford and G. Marcus (eds) *Writing Culture: The Poetics and Politics of Ethnography*. Berkeley: University of California Press, pp. 234–62.

Rahman, A. (1993) *People's Self-development: Perspectives on Participatory Action Research*. London: Zed Books.

Rahnema, M. (1992) 'Participation'. In W. Sachs (ed.) *The Development Dictionary: A Guide to Knowledge as Power*. London: Zed Books, pp. 116–32.

Rajak, D. (2011) *In Good Company: An Anatomy of Corporate Social Responsibility*. Stanford, CA: Stanford University Press.

Rajak, D. and R.L. Stirrat (2011) 'Parochial cosmopolitanism and the power of nostalgia.' In D. Mosse (ed.) *Adventures in Aidland: The Anthropology of Professional International Development*. Oxford: Berghahn, pp. 161–77.

Redfield, P. (2012) 'Bioexpectations: life technologies as humanitarian goods', *Public Culture* 24(1): 157–84.

Reiter, R. (ed.) (1975) *Toward an Anthropology of Women*. New York: Monthly Review.

Rew, A. (1985) 'The organizational connection: multi-disciplinary practice and anthropological theory'. In R. Grillo and A. Rew (eds) *Social Anthropology and Development Policy*, ASA Monographs 23. London: Tavistock, pp. 185–98.

Rhoades, R.E. (1984) *Breaking New Ground: Agricultural Anthropology*. Lima: International Potato Center.

Richards, A. (1939) *Land, Labour and Diet in Northern Rhodesia*. London: Oxford University Press.

Richards, P. (1993) 'Cultivation: knowledge or performance?' In M. Hobart (ed.) *An Anthropological Critique of Development: The Growth of Ignorance*. London: Routledge, pp. 61–79.

Riddell, R. (2008) *Does Foreign Aid Really Work?* Oxford: Oxford University Press.

Robertson, A.F. (1984) *The People and the State: An Anthropology of Planned Development*. Cambridge: Cambridge University Press.

Robbins, J. (2004) *Becoming Sinners: Christiantity and Moral Torment in a Papua New Guinea Society*. Berkeley: University of California Press.

Roe, E. (1991) 'Development narratives, or making the best of blueprint development', *World Development* 19(4): 287–300.

Rogers, B. (1980) *The Domestication of Women: Discrimination in Developing Societies*. London: Kogan Page.

Rogers, D. (2012) 'The materiality of the corporation: oil, gas, and corporate social technologies in the remaking of a Russian region', *American Ethnologist* 39: 284–96.

Roodman, D. and J.Morduch (2014) 'The impact of microcredit on the poor in Bangladesh: revisiting the evidence', *Journal of Development Studies* 50(4): 583–604.

Rosaldo, M. and L. Lamphere (eds) (1974) *Woman, Culture and Society*. Stanford, CA: Stanford University Press.

Rostow, W.W. (1960a) *The Process of Economic Growth*, 2nd edn. London: Clarendon.

Rostow, W.W. (1960b) *The Stages of Economic Growth: A Non-communist Manifesto.* Cambridge: Cambridge University Press.

Rottenburg, R. (2009) *Far-fetched Facts: A Parable of Development Aid.* Cambridge, MA: Massachusetts Institute of Technology.

Rozario, S. (1992) *Purity and Communal Boundaries: Women and Social Change in a Bangladeshi Village.* London: Zed Books.

Rylko-Bauer, B., M. Singer and J. Van Willegen (2006) 'Reclaiming applied anthropology: its past, present and future', *American Anthropologist* 108(1): 178–90.

Sachs, W. (ed.) (1992) *The Development Dictionary: A Guide to Knowledge as Power.* London: Zed Books.

Sacks, K. (1975) 'Engels revisited: women, the organisation of production and private property'. In R. Reiter (ed.) *Toward an Anthropology of Women.* New York; Monthly Review, pp. 211–35.

Said, E. (1978) *Orientalism.* Harmondsworth: Penguin.

Salisbury, R. (1977) 'A prism of perceptions: the James Bay hydro electricity project'. In S. Wallman (ed.) *Perceptions of Development.* Cambridge: Cambridge University Press.

Sanchez, A. (2012) 'Questioning success: dispossession and the criminal entrepreneur in urban India', *Critique of Anthropology* 32(4): 435–57.

Sawyer, S. (2005) *Crude Chronicles: Indigenous Politics, Multinational Oil and Neoliberalism in Ecuador.* Durham, NC: Duke University Press.

Schapera, I. (1947) *Migration and Tribal Life.* Oxford: Oxford University Press.

Schaumberg, H. (2008) 'Taking sides in the oilfields: for a politically engaged anthropology'. In H. Armbruster and A. Laerke (eds) *Taking Sides: Ethics, Politics and Fieldwork in Anthropology.* Oxford: Berghahn, pp. 199–216.

Scherz, C. (2014) *Having People, Having Heart: Charity, Sustainable Development, and Problems of Dependence in Central Uganda.* Chicago: University of Chicago Press.

Schuller, M. and D. Lewis (2014) 'Anthropology of NGOs'. In J. Jackson (ed.) *Oxford Bibliographies in Anthropology.* New York: Oxford University Press.

Schwegler, T. (2008) 'Take it from the top (down)? Rethinking neo-liberalism and political hierarchy in Mexico', *American Ethnologist* 35(4): 682–700.

Scoones, I. and J. Thompson (1993) *Challenging the Populist Perspective: Rural Peoples' Knowledge, Agricultural Research and Extension Practice.* Institute of Development Studies Discussion Paper 332. Sussex: Institute of Development Studies.

Scoones, I. and J. Thompson (1994) *Beyond Farmer First: Rural Peoples' Knowledge, Agricultural Research and Extension Practice.* London: Intermediate Technology Publications.

Scoones, I. and J. Thompson (eds) (2009) *Farmer First Revisited: Innovation for Agricultural Research and Development.* Rugby: Practical Action Publishing.

Scott, J. (1998) *Seeing Like a State: How Certain Schemes to Improve the Human Condition have Failed*. New Haven, CT: Yale University Press.

Scudder, T. (1980) 'Policy implications of complusory relocation in river basin development projects'. In M. Cernea and P.B. Hammond (eds) *Projects for Rural Development: The Human Dimension*. Baltimore, MD: Johns Hopkins University Press.

Seddon, D. (1993) 'Anthropology and appraisal: the preparation of two IFAD pastoral development projects in Niger and Mali'. In J. Pottier (ed.) *Practising Development: Social Science Perspectives*. London: Routledge, pp. 71–110.

Sen, A. (1981) *Poverty and Famines: An Essay on Entitlement and Deprivation*. Oxford: Oxford University Press.

Sen, G. and C. Grown (1987) *Development Crises and Alternative Visions: Third World Women's Perspectives*. New York: Monthly Review.

Shah, A. (2013) 'The intimacy of insurgency: beyond coercion, greed or grievance in Maoist India', *Economy and Society* 42(3): 480–506.

Shever, E. (2010) 'Engendering the company: corporate personhood and the "face" of an oil company in metropolitan Buenos Aires', *PoLAR – Political and Legal Anthropology Review* 33(1): 26–45.

Shore, C. and S. Wright (1997) 'Policy: a new field of anthropology'. In C. Shore and S. Wright (eds) *Anthropology of Policy: Critical Perspectives on Governance and Power*. London: Routledge.

Shore, C. and S. Wright (eds) (2003) *Anthropology of Policy: Perspectives on Governance and Power*. Abingdon: Routledge.

Shore, C., S. Wright and D. Però (eds) (2011) *Policy Worlds: Anthropology and the Analysis of Contemporary Power*. Oxford: Berghahn.

Sillitoe, P. (ed.) (2007) *Local Science vs. Global Science: Approaches to Indigenous Knowledge in International Development*. Oxford: Berghahn.

Sittón, S.N. (2011) 'The role of anthropology in the changes and challenges of 21st century', Malinowski lecture presented to the 71st Annual Meeting of the Society for Applied Anthropology (SfAA), Seattle, WA, March–April.

Smith, D.J. (2007 *A Culture of Corruption: Everyday Deception and Popular Discontent in Nigeria*. Princeton, NJ: Princeton University Press.

Smith, J.H. (2008) *Bewitching Development: Witchcraft and the Reinvention of Development in Neo-Liberal Kenya*. Chicago: University of Chicago Press.

Smith, J. and Helfgott, F. (2010) 'Flexibility or exploitation: corporate social responsibility and the perils of universalization', *Anthropology Today* 26(3): 20–23.

Sobhan, R. (1989) 'Bangladesh and the world economic system: the crisis of external dependence'. In H. Alavi and J. Harriss (eds) *Sociology of 'Developing Societies': South Asia*. London: Macmillan.

Soederberg, S. (2004) *The Politics of the New International Financial Architecture: Reimposing Neoliberal Domination in the Global South*. London: Zed Books.

Spencer, J. (1989) 'Anthropology as a kind of writing', *Man: The Journal of the Royal Anthropological Institute* 24(1): 145–64.

Standing, G. (2011) *The Precariat: The New Dangerous Class*. London: Bloomsbury.

Staudt, K. (ed.) (1990) *Women, International Development and Politics: The Bureaucratic Mire*. Philadelphia, PA: Temple University Press.

Staudt, K. (1991) *Managing Development: State, Society, and International Contexts*. Newbury Park, CA: Sage.

Staudt, K. (1998) *Policy, Politics and Gender: Women Gaining Ground*. West Hartford, CT: Kumarian.

Stirrat, R.L. and H. Henkel (1997) 'The development gift: the problem of reciprocity and the NGO world', *Annals of the American Academy of Political and Social Science* 554: 66–80.

Strümpell, C. (2014a) 'The making of a working class in West-Odisha.' Paper presented at Humboldt-Universität zu Berlin, Philosophische Fakultät III, Institut für Asien-und Afrikawissenschaften, Seminar für Südasien-Studien.

Strümpell, C. (2014b) 'The politics of dispossession in an Odishan steel town', *Contributions to Indian Sociology* 48(1): 45–72.

Sutcliffe, B. (2005) *A Converging or Diverging World?* DESA Working Paper Series ST/ESA/2005/DWP/2. New York: UN Department of Economic and Social Affairs.

Taussig, M. (1980) *The Devil and Commodity Fetishism in South America*. Chapel Hill, SC: South Carolina Press.

Tett, G. (2010) *Fool's Gold: How Unrestrained Greed Corrupted a Dream, Shattered Global Markets and Released a Catastrophe*. London: Abacus.

Thin, N. (2009) 'Why anthropology can ill-afford to ignore wellbeing'. In G. Matthews and C. Izquierdo (eds) *Pursuits of Happiness: Well-being in Anthropological Perspective*. Oxford: Berghahn.

Turner, J. (1969) 'Uncontrolled urban settlements: problems and policies'. In G. Breese (ed.) *The City in Newly Developing Countries: Readings on Urbanism and Urbanisation*. New York: Prentice Hall, pp. 507–31.

UNDP (United Nations Development Programme) (2013) *Human Development Report 2013 – The Rise of the South: Human Progress in a Diverse World*. New York: UNDP.

van Willigen, John (1993) *Applied Anthropology: An Introduction*, rev. edn. Westport CT: Bergin and Garvey.

Vatuk, S. (1972) *Kinship and Urbanisation: White Collar Migrants in North India*. London: University of California Press.

Walker, K. Le M. (2008) 'Neoliberalism on the ground in rural India: predatory growth, agrarian crisis, internal colonization, and the intensification of class struggle', *Journal of Peasant Studies* 35(4): 557–620.

Wallace, T., F. Porter and M. Ralph-Bowman (2013) *A Perfect Storm: Aid, NGOs and the Realities of Women's Lives*. Rugby: Practical Action Publishing.

Wallerstein, I. (1974) *The Modern World System: Capitalist Agriculture and the Origins of the European World Economy in the Sixteenth Century.* New York: Academic Press.

Warren, B. (1980) *Imperialism: Pioneer of Capitalism.* London: Verso.

Watanabe, C. (2013) 'Past loss as future? The politics of temporality and the "non-religious" by a Japanese NGO in Burma/Myanmar', *POLAR – Political and Legal Anthropology Review* 36(1): 75–98.

Weiner, A. (1976) *Women of Value, Men of Renown.* Austin: University of Texas Press.

Welker, M. (2009 '"Corporate security begins in the community": mining, the CSR industry and environmental advocacy in Indonesia', *Cultural Anthropology* 24(1): 142–79.

Welker, M. (2012) 'The Green Revolution's ghost: unruly subjects of participatory development in rural Indonesia', *American Ethnologist* 39(2): 389–406.

Welker, M. (2014) *Enacting the Corporation: An American Mining Firm in Post-Authoritarian Indonesia.* Berkeley: University of California Press.

Welker, M., D.J. Partridge and R. Hardin (eds) (2011) 'Corporate lives: new perspectives on the social life of the corporate form'. Introduction to Supplement 3, *Current Anthropology*, 52(S3): S3–S16.

White, S. (1992) *Arguing with the Crocodile: Gender and Class in Bangladesh.* London: Zed Books.

Whitehead, A. (1981) '"I'm hungry Mum": the politics of domestic budgeting'. In K. Young, C. Wolkowitz and R. McCullagh (eds) *Of Marriage and the Market: Women's Subordination Internationally and its Lessons.* London: Routledge and Kegan Paul, pp. 93–117.

Wilson, G. (1941) *An Essay on the Economics of Detribalisation of Northern Rhodesia, Part I.* Rhodes-Livingstone Papers 5. Livingstone: Rhodes-Livingstone Institute.

Wilson, G. (1942) *An Essay on the Economics of Detribalisation of Northern Rhodesia, Part II.* Rhodes-Livingstone Papers 6. Livingstone: Rhodes-Livingstone Institute.

Win, E.J. (2007): 'Not very poor, powerless or pregnant: the African woman forgotten by development', *IDES Bulletin* 35: 61–64.

Wolf, E. (1964) *Anthropology.* London: Prentice-Hall.

Wolf, E. (1982) *Europe and the People without History.* Berkeley: University of California Press.

Wood, A. (1950) *The Ground-Nut Affair.* London: Bodley Head.

Worby, E. (1984) 'The politics of dispossession: livestock development policy and the transformation of property relations in Botswana'. Unpublished MA thesis, Department of Anthropology, McGill University, Montreal.

World Bank (2001) *World Development Report 2000/2001: Attacking Poverty.* New York: Oxford University Press.

Worsley, P. (1957) *The Trumpet Shall Sound: A Study of 'Cargo Cults' in Melanesia.* London: MacGibbon and Kee.

Worsley, P. (1984) *The Three Worlds: Culture and World Development.* London: Weidenfeld and Nicolson.

Wright, S. (1995) 'Anthropology: still the uncomfortable discipline?' In A. Ahmed and C. Shore (eds) *The Future of Anthropology: Its Relevance to the Contemporary World*. London: Athlone.

Wright, S. (2006) 'Machetes into a jungle? A history of anthropology in policy and practice 1981–2000'. In S. Pink (ed.) *Applications of Anthropology: Professional Anthropology in the Twenty-first Century*. Oxford: Berghahn, pp. 27–54.

Yarrow, T. (2011) 'Maintaining independence: the moral ambiguities of personal relations among Ghanaian development workers.' In H. Hindman and A.-M. Fechter (eds) *Inside the Everyday Lives of Aidworkers*. West Hartford, CT: Kumarian, pp. 41–58.

Yeh, E.T. (2013) *Taming Tibet: Landscape Transformation and the Gift of Chinese Development*. Ithaca, NY: Cornell University Press.

Yessanova, S. (2012) 'The Tengiz oil enclave: labour business and the state', *PoLAR – Political and Legal Anthropology Review* 35(1): 94–114.

Zalik, A. (2004) 'The Niger Delta: "petro violence" and "partnership development"', *Review of African Political Economy* 31(101): 401–24.

Index

Compiled by Sue Carlton